AF289136

sona nanpa Paluli

paw'II' mI'QeD

kepeken toki tu wan:
sitelen pona, toki pona, toki Sinan

nutlhej wej Hol:
SI'telenpo'na Hol, to'qIpo'na Hol, tlhIngan Hol

Martin Erik Horn / jan Masin Elki / mI'mey Dop Qaw'wI'

Martin Erik Horn / mI'mey Dop Qaw'wI' /
jan Masin Elki Pi Ilo Kalama Pi Palisa Lawa
Lilienthalpark / lI'lIymey ngech yotlh /
ma kasi pi jan tawa sewi nanpa wan
Berlin / berlIn / ma tomo Pelin
Germany / DoyIchlan / ma Tosi

Email:
toki-pona@clifford-algebra.de

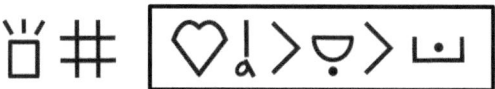

sona nanpa Paluli

paw'lI' mI'QeD

kepeken toki tu wan:
sitelen pona, toki pona, toki Sinan

nutlhej wej Hol:
SI'telenpo'na Hol, to'qlpo'na Hol, tlhIngan Hol

Martin Erik Horn
jan Masin Elki Pi Ilo Kalama Pi Palisa Lawa
mI'mey Dop Qaw'wI'

Die vorliegende Fassung ist eine Übersetzung des Buches
„Pauli Algebra – sona nanpa Paluli", ursprünglich erschienen bei
BoD unter der ISBN 978-3-7597-5001-3 (c) 2024 Martin Erik Horn,
auf Toki Pona, englisch und deutsch.

Bibliographische Information der Deutschen Nationalbibliothek:

Die Deutsche Nationalbibliothek verzeichnet diese Publikation
in der Deutschen Nationalbiographie; detaillierte bibliographische
Daten sind im Internet über

http://dnb.dnb.de

abrufbar.

Herstellung und Verlag:
BoD - Books on Demand, Norderstedt

ISBN: 978-3-7597-6759-2

**sina ken ala toki e ijo, lon toki pona
la sina sona pona ala e ona.**

Hol nap DajatlhtaHvIS DaQIjbe'chugh
Dayajlaw'pu'be'.

o lukin e ni / naDev yllegh:

Sonja Lang: Toki Pona.
The Language of Good.
The official Toki Pona book pu.
(Quotation attributed to
Albert Einstein or
Richard Feynman)

**sina ken ala toki e ijo, lon toki pona
la mi sona pona ala e ona.**

Hol nap DajatlhtaHvIS DaQIjbe'chugh
wIyajlaHbe' neH.

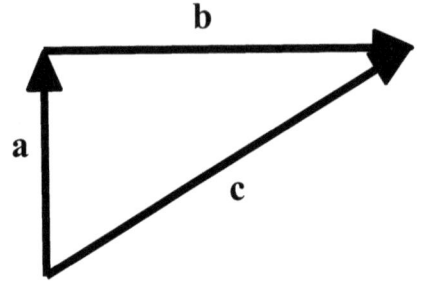

↓〉⊓⌐∟⌣∥１

ni li selo pi linja tu wan.

naDev ra'Duch tu'lu'.

∽〉Ɛ≫⌐·∥１

ona li jo e poka tu wan.

wej reD ghaj.

⌐·♯１＋⌐·♯∥〉ᵛ
⌐·♯∥１〉V

poka nanpa wan en poka nanpa tu li lili.
poka nanpa tu wan li suli.

wII reD wa'DIch, reD cha'DIch je.
tIq reD wejDIch.

6

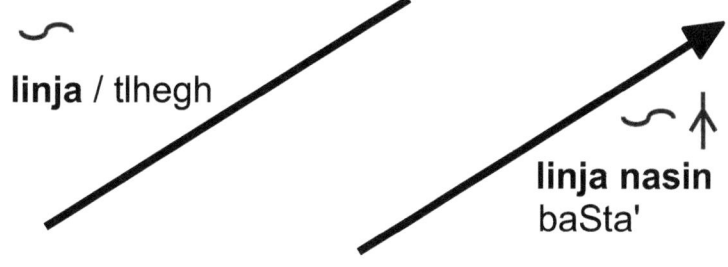

linja / tlhegh

linja nasin
baSta'

poka li jo e nasin.

ona li linja nasin.

lurghmey ghaj reDmey. baSta'mey bIH.

mi sitelen e linja nasin, kepeken sitelen wawa.

baSta'mey DIghItlhchugh ngutlhmey pI' DIlo'.

linja nasin / baSta'mey: a

b

c

K)⌒↑Ⅲ>Ⅲ

ken la linja nasin mute li mute.

baSta'mey boqlaH baSta'mey.

$$a + b = c$$

⌒↑♯1>Ⅲ≫⌒↑♯Ⅱ)
↓>=≫⌒↑♯Ⅱ1

linja nasin nanpa wan li mute e linja nasin nanpa tu la ni li sama e linja nasin nanpa tu wan.

baSta' cha'DIch boq baSta' wa'DIch;
chen baSta' wejDIch.

$$K) \frown \Uparrow ||| \rangle \overset{\circ}{\circ}{}_{\circ}$$

ken la linja nasin mute li kulupu.

'oplogh baSta'mey boq'eghlaH baSta'mey.

$$K) \frown \Uparrow \rangle \overset{\circ}{\circ}{}_{\circ} \gg \backsim \varpi =$$

ken la linja nasin li kulupu e ona sama.

'oplogh baSta'mey nIb boq'eghlaH baSta'mey.

$$a\ a = a^2$$

$$\text{ΛΛΛ} \llcorner \breve{\square} \# \rangle \breve{\circ} :$$
$$\frown \Uparrow \rangle \overset{\circ}{\circ}{}_{\circ} \gg \frown \Uparrow =) \Downarrow \rangle = \gg \boxdot$$

monsuta pi sona nanpa li toki:
linja nasin li kulupu e linja nasin sama
la ni li sama e leko.

jatlh mI'QeD veqlargh:
wa'logh baSta' nIb boq'egh baSta'; chen meyrl'.

X

ala !

ghobe' !

⌒↑〉⸛≫⌒↑＝〉↓〉＝≫#

**linja nasin li kulupu e linja nasin sama
la ni li sama e nanpa.**

wa'logh baSta' nlb boq'egh baSta'; chen ml'.

⭕ 𖨂 :

ijo pana / ghantoH:

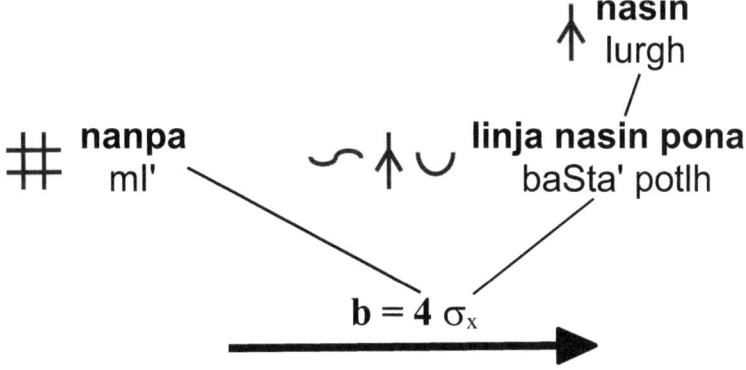

nasin
lurgh

nanpa
mI'

linja nasin pona
baSta' potlh

$b = 4\ \sigma_x$

linja nasin ni li sama e noka tu tu tawa supa.

loS qam chuq SaS tu'lu'; chen baSta'vam.

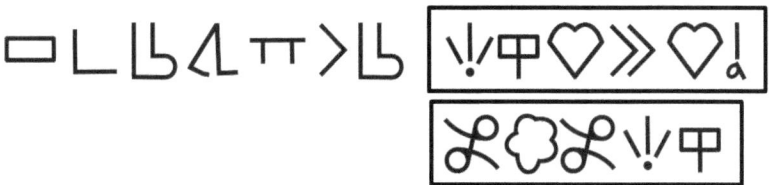

nimi pi noka tawa supa li noka Sipepa Ekesi.

baSta' potlh SaS 'oH qam chuqvam pong'e'.

$$4 \cdot 4 \qquad \sigma_x \, \sigma_x$$

$$b^2 = b \, b = (4 \, \sigma_x)^2 = (4 \, \sigma_x)(4 \, \sigma_x) = 16 \, \sigma_x^2 = 16$$

⌒↑↓〉⸛》ᴖ◦＝)‖‖〉⸛》‖‖
¡)Ⳁ1⊲ᴛᴛ〉⸛》Ⳁ1⊲ᴛᴛ

linja nasin ni li kulupu e ona sama la tu tu li kulupu e tu tu. kin la noka wan tawa supa li kulupu e noka wan tawa supa.

wa'logh baSta' nIb boq'eghchugh baSta'vam; chen; loSlogh boq'egh loS 'ej wa'logh qam chuq SaS boq'egh wa' qam chuq SaS.

‖‖〉⸛》‖‖)↓〉＝》∩∩∩1

tu tu li kulupu e tu tu la ni li sama e luka luka luka wan.

loSlogh boq'egh loS; chen wa'maH jav.

▭Ⳑ ⳑ ✕〉Ⳑ   〡¡⾌♡》♡¡
                    &⅋ᘕᑎ☮⾌

nimi pi noka ante li noka Sipepa Jenki.

baSta' potlh chong 'oH latlh qam chuq pong'e'.

ᒪ1>ᵒ₀ᵒ ≫ ᒪ1 =)↓> = ≫?

**noka wan li kulupu e noka wan sama la
ni li sama e seme?**

wa'logh qam chuq nIb boq'egh wa' qam chuq;
chen nuq ?

θᵕ#1:

lawa pona nanpa wan:

mI'QeD chut wa'DIch:

$$\sigma_x{}^2 = 1$$

$$\sigma_y{}^2 = 1$$

ᒪ1>ᵒ₀ᵒ ≫ ᵕᵒ =)↓> = ≫1

noka wan li kulupu e ona sama la ni li sama e wan.

wa'logh chuqvam boq'egh wa' qam chuq; chen wa'.

13

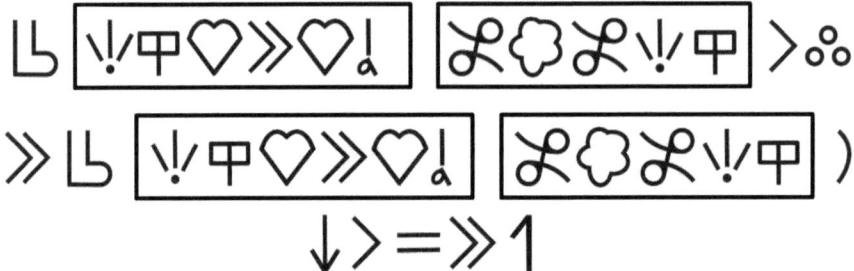

noka Sipepa Ekesi li kulupu e noka Sipepa Ekesi la ni li sama e wan.

wa'logh baSta' potlh SaS boq'egh wa' baSta' potlh SaS; chen wa'.

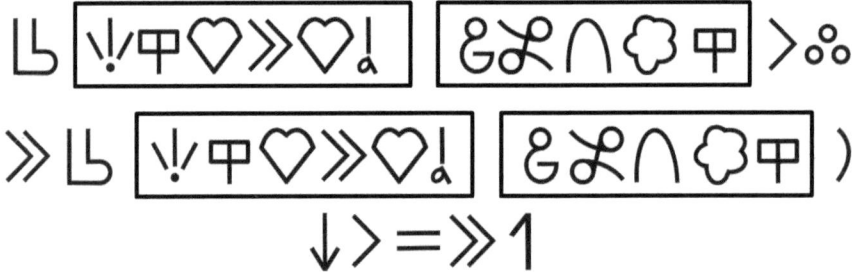

noka Sipepa Jenki li kulupu e noka Sipepa Jenki la ni li sama e wan.

wa'logh baSta' potlh chong boq'egh wa' baSta' potlh chong; chen wa'.

$$b^2 = (4\,\sigma_x)^2 = 16\,\sigma_x^2 = 16$$

noka tu tu li kulupu e ona sama la ni li sama e luka luka luka wan.

loSlogh chuqvam boq'egh loS qam chuq; chen wa'maH jav.

noka Sipepa Ekesi tu tu li kulupu e noka Sipepa Ekesi tu tu la ni li sama e luka luka luka wan.

loSlogh baSta' potlh SaS boq'egh loS baSta' potlh SaS; chen wa'maH jav.

sona nanpa li lon.

yln ml'QeD.

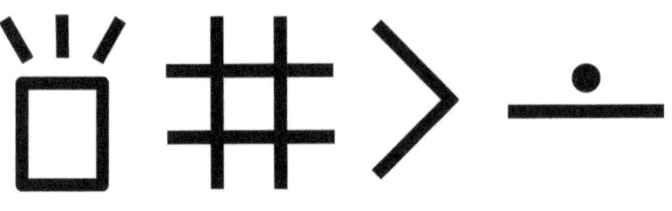

jan Pisakojasa li lon a!

yln peytlharghngongraS !

ᔆ ⟩⟩ Ω | ♡⌣–ᵎ-↓Ƙ♡ℰ↓-ᵎ-↓ |

mi open alasa e jan Pisakojasa.

DaH peytlharghngongraS wInej.

Ω | ♡⌣–ᵎ-↓Ƙ♡ℰ↓-ᵎ-↓ | ⟩♡ ⟩⟩ ▣

jan Pisakojasa li olin e leko.

meyrI'mey muSHa' peytlharghngongraS.

$$c^2 = (a + b)^2$$

⌣·#‖1⟩°₀°⟩⟩ᵔᵒ=

poka nanpa tu wan li kulupu e ona sama.

wa'logh reDvam boq'egh reD wejDIch.

⌣·#1⟩Ⲓ‖‖⟩⟩⌣·#‖)↓⟩°₀°⟩⟩ᵔᵒ=

**poka nanpa wan li pini mute e poka nanpa tu
la ni li kulupu e ona sama.**

reD cha'DIch boq reD wa'DIch; chen gher'ID ru'.
vaj wa'logh gher'IDvam boq'egh gher'IDvam.

$$c^2 = (a + b)^2 = (a + b)\,(a + b)$$

$$⊔\cdot\#1+⊔\cdot\#\,\|\,\rangle\overset{\circ}{\circ}$$
$$\gg⊔\cdot\#1+⊔\cdot\#\,\|$$

poka nanpa wan en poka nanpa tu li kulupu e poka nanpa wan e poka nanpa tu.

reD cha'DIch boqta' reD wa'DIch boq'egh reD cha'DIch boqta' reD wa'DIch.

$$\text{WW}∟ö\#\rangle ö: \quad ↓\rangle = \gg θ ♀$$

monsuta pi sona nanpa li toki: ni li sama e lawa tonsi.

jatlh mI'QeD veqlargh: cha' 'ay'vaD mI'QeD chut 'oH wItte'vam'e'.

$$(a + b)^2 = (a + b)\,(a + b)$$
$$= a^2 + 2\,ab + b^2$$

$$+↓\rangle\cap \qquad ↓\rangle \div X$$

taso ni li ike !   ni li lon ala !

'ach muj mI'QeD veqlargh ! mItbe' wItte'vam !

X ↓≻=X≫θ♀

ala ! ni li sama ala e lawa tonsi.

ghobe'; cha' 'ay'vaD ml'QeD chut rurbe' wltte'vam.

$\text{P} \wedge \& \gg \% \| \|$

mi kama jo e kipisi tu tu.

loS 'ay' DISuq.

o lukin e sitelen linja !

naQjejHommey tlegh !

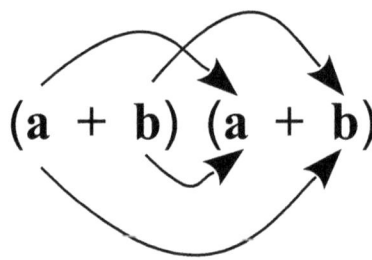

$$(a + b) \ (a + b)$$

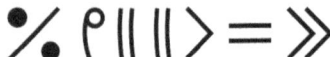

kipisi mi tu tu li sama e ...

$$a\,a = a^2$$
$$a\,b$$
$$b\,a$$
$$b\,b = b^2$$

... blH loS 'ay'meymaj'e'.

$$c^2 = (a + b)^2 = (a + b)\,(a + b) = a^2 + a\,b + b\,a + b^2$$

⊔·♯1＋⊔·♯‖＞°ₒ
≫⊔·♯1≫⊔·♯‖
)↓＞＝≫⊔·♯1∟⊔·♯1
≫⊔·♯1∟⊔·♯‖
≫⊔·♯‖∟⊔·♯1
≫⊔·♯‖∟⊔·♯‖

**poka nanpa wan en poka nanpa tu li kulupu
e poka nanpa wan e poka nanpa tu
la ni li sama e poka nanpa wan pi poka nanpa wan
e poka nanpa wan pi poka nanpa tu
e poka nanpa tu pi poka nanpa wan
e poka nanpa tu pi poka nanpa tu.**

wa' reD cha'DIch boq wa' reD wa'DIch; 'oplogh
boq'egh; wa' reD cha'DIch boq wa' reD wa'DIch;
vaj chen;
wa'logh reD wa'DIch boq'egh wa' reD wa'DIch
'ej wa'logh reD cha'DIch boq'egh wa' reD wa'DIch
'ej wa'logh reD wa'DIch boq'egh wa' reD cha'DIch
'ej wa'logh reD cha'DIch boq'egh wa' reD cha'DIch.

⊔·#1+⊔·#‖>ᘓ≫↑⊣)Ω

♡⊔-ˡ-↓ⱪⴹᘓↄ-ˡ-↓ >⋀≫ᴛᴛ℧

**poka nanpa wan en poka nanpa tu li jo e nasin
taso la jan Pisakojasa li kama e supa musi.**

leDchugh reD wa'DIch, reD cha'DIch je,
much yaH 'el peytlharghngongraS.

↓>☺ː

ni li toki pona / to'qIpo'na Hol 'oH:

ᘓ≫↑＝   **jo e nasin sama**
                lurgh nIb ghaj = Don

ᘓ╳≫↑＝   **jo ala e nasin sama**
                lurgh nIb ghajbe' = DonHa'

ᘓ≫↑╳   **jo e nasin ante**
                lurgh pIm ghaj

ᘓ≫↑-ˡ-   **jo e nasin sin**
                latlh lurgh ghaj

ᘓ≫↑⊣   **jo e nasin taso**
                lurgh mob ghaj
                = nIteb lurghvam ghaj = leD

22

⊙✕∶

lukin ante / DIDay: $\qquad c^2 = a^2 + a\,b + b\,a + b^2$

Q ♡ ⌣∙ ⌐′⌐↓ K ♡ & ↓ ⌐′⌐↓ ⟩ ♡̈ ∶

jan Pisakojasa li toki:
jatlh peytlharghngongraS: $\qquad c^2 = a^2 \qquad\qquad + b^2$

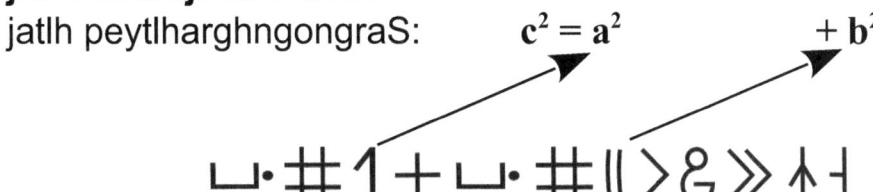

poka nanpa wan en poka nanpa tu li jo e nasin taso.
leD reD wa'DIch, reD cha'DIch je.

♡̈ I ∶

sona pini / gher'ID ngaD:

$$c^2 = a^2 + \underbrace{a\,b + b\,a}_{0} + b^2$$

$$\Rightarrow \qquad a\,b + b\,a = 0$$

$$\Rightarrow \qquad a\,b = -\,b\,a$$

⊡ ∿ **sitelen jasima**
maQ Dop

23

$$a\,b = -\,b\,a$$

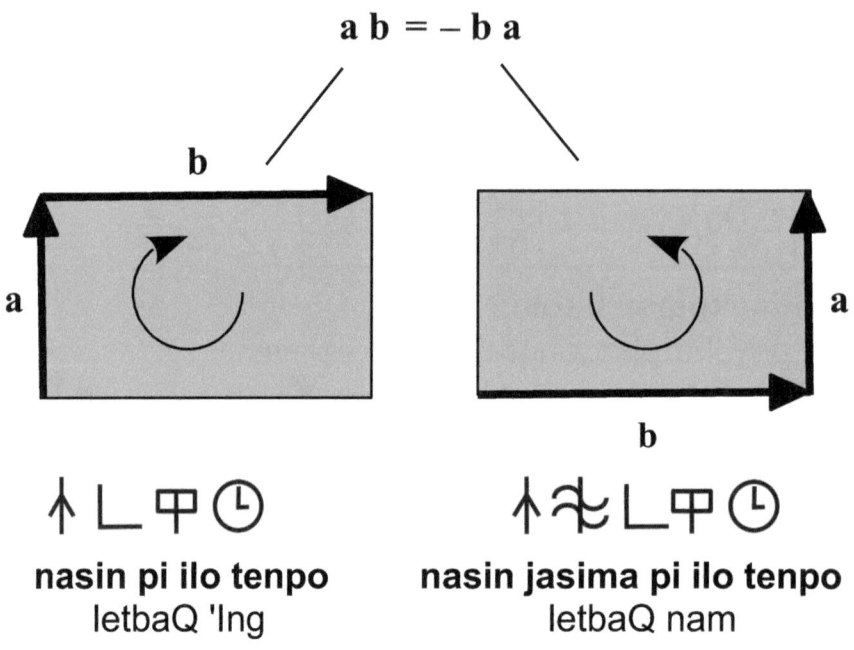

𝔸𝐋ꟼ🕐

nasin pi ilo tenpo
letbaQ 'Ing

𝔸⚘𝐋ꟼ🕐

nasin jasima pi ilo tenpo
letbaQ nam

mi ante e nasin poka la sitelen jasima li kama.

reDmey Dotlh cho' wIyoymoHchugh, nargh maQ Dop.

mi jo e sitelen jasima la nasin pi selo li ante.

maQ Dop wIghajchugh, mI'QeD tu'qom cho' tamlu'.

ꓑ⊔𖢕》〜⼂⌣

mi open kepeken e linja nasin pona.

DaH basta'mey potlh DIlo'choH.

$a = \sigma_y$

$b = \sigma_x$ $\Rightarrow$ $a\,b = \sigma_y\,\sigma_x$

$b\,a = \sigma_x\,\sigma_y$

ŎＩ：

sona pini / gher'ID ngaD:

$a\,b = -\,b\,a$ $\Rightarrow$ $\sigma_y\,\sigma_x = -\,\sigma_x\,\sigma_y$

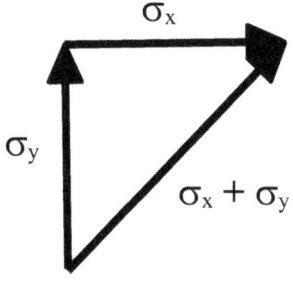

σ_x

σ_y

$\sigma_x + \sigma_y$

Ｙ：

anu / qoj: $b\,a = -\,a\,b$ $\Rightarrow$ $\sigma_x\,\sigma_y = -\,\sigma_y\,\sigma_x$

θ ⌣ # ‖ :

lawa pona nanpa tu:

mI'QeD chut cha'DIch:

$$\sigma_x \sigma_y = -\sigma_y \sigma_x$$

↓ > ▣

ni li leko.

naDev meyrI'mey bIHtaH.

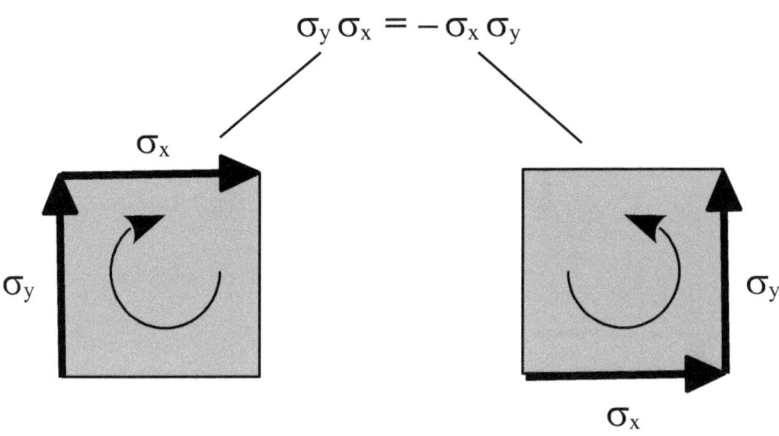

$\sigma_y \sigma_x = -\sigma_x \sigma_y$

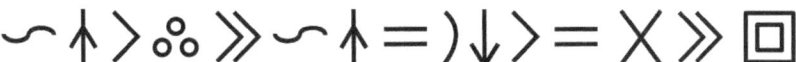

monsuta pi sona nanpa li pilin ike mute !

'IQqu' mI'QeD veqlargh.

**linja nasin li kulupu e linja nasin sama
la ni li sama ala e leko.**

wa'logh baSta' nIb boq'egh baSta'; chenbe' meyrl'.

**taso linja nasin pona li kulupu e linja nasin pona
ante la ni li sama e leko …**

'ach wa'logh baSta' potlh pIm boq'egh baSta' potlh;
chen meyrl' …

… anu e selo pi linja nasin taso.

… letbaQ joq.

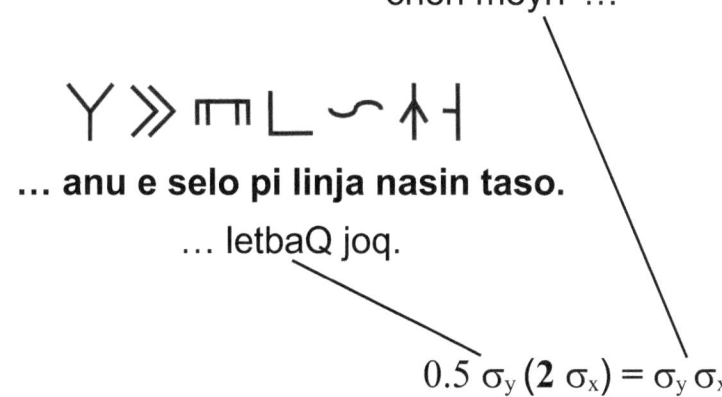

$$0.5\,\sigma_y\,(\mathbf{2}\,\sigma_x) = \sigma_y\,\sigma_x$$

回 ﹥ 中 ⌣ 凵 ⺆ ⼞ 井

leko li ilo pona a pi sona nanpa.

ml'QeD Somml'mey Dunqu' bIH meyrl'mey'e'.

＋ 回 ﹥ 𝟪 ≫ ⼞ 冎

en leko li jo e sona len.

'ej pegh lughaj meyrl'mey.

回 回 ﹥ 𝟨

leko leko li nasa.

taQ meyrl'mey meyrl'mey.

ona li nasa a li nasa nasa a !

taQqu' 'ej jumqu' !

◨ ◨ :

leko leko / meyrl'mey meyrl'mey:

$$- \sigma_x \sigma_y$$

$$(\sigma_x \sigma_y)^2 = (\sigma_x \sigma_y)(\sigma_x \sigma_y) = \sigma_x \overbrace{\sigma_y \sigma_x} \sigma_y$$

$$= - \sigma_x \sigma_x \sigma_y \sigma_y$$

$$= - \sigma_x^2 \sigma_y^2$$

$$= - 1$$

◨ ↓ ∟ ◨ ◡ ⟩ = ≫ 1 ꩜

leko ni pi leko pona li sama e wan jasima.

meyrl' potlh meyrl'vam tu'lu'; chen wa' Dop.

◨ ◨ ⟩ ♯ ꩜

leko leko li nanpa jasima.

ml'mey Dop blH meyrl'mey meyrl'mey'e'.

◷ Ⅰ :

sona pini / gher'ID ngaD:

◨ ⟩ = ✕ ≫ ♯ ∻

leko li sama ala e nanpa lon.

meyrl' tu'lu', chenbe' ml' ruj. ml' ruj nelbe' meyrl'.

29

leko li sama e ijo nasa.

vay' jal nel meyrl'. jalwl'mey blH meyrl'mey'e'.

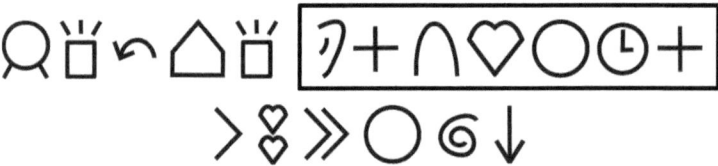

**jan sona tan tomo sona Kenpite li olin
e ijo nasa ni.**

qam blQtlq QI retlhDaq DuSaQ quvDaq
vay'vam jal luparHa'qu' tejpu'.

o lukin e ni / naDev yllegh:

Stephen Gull, Anthony Lasenby, Chris Doran:
Imaginary Numbers are not Real – The Geometric
Algebra of Spacetime. Foundation of Physics,
Vol. 23, No. 9 (1993), pp. 1175 – 1201.

ona li toki / jatlh:

> “We have now reached the point which is liable to cause the greatest intellectual shock.“

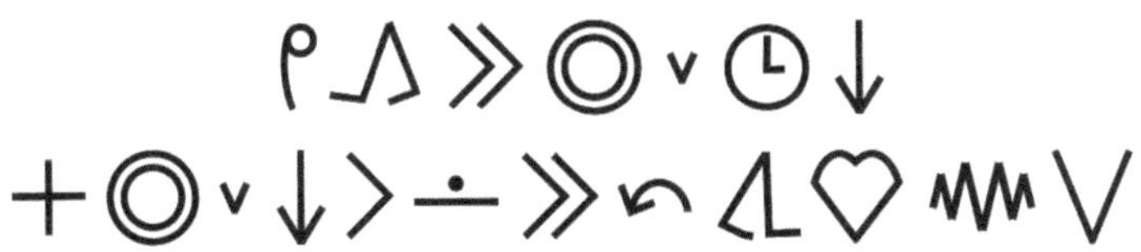

mi kama e sike lili, tenpo ni.
en sike lili ni li lon e tan, tawa pilin monsuta suli.

DaH DaqHom wISIch. 'ej nughIjmoHqu' DaqHomvam.
wIHajqu'choH. maDuqqu'.

> "… we get … exactly the algebra of the Pauli spin matrices used in the quantum mechanics of spin-1/2 particles!"

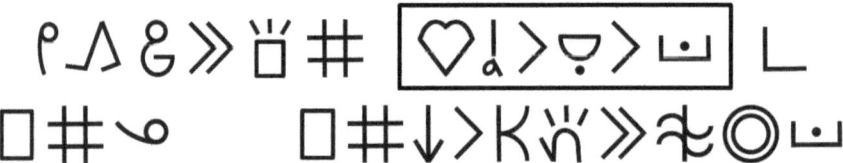

mi kama jo e sona nanpa Paluli pi lipu nanpa ona. lipu nanpa ni li ken pana e jasima sike insa.

ml'mey tlhatmeyDaj ghajbogh paw'll' ml'QeD wlSuq.
jlrbogh qoDmey Del ml'mey tlhatmeyvam
qoj qoD DIngta'ghach luDel.

ᕈႮ≫▢#↓ᚼ☱L◯ˇ

mi kepeken e lipu nanpa ni, tawa sona pi ijo lili.

pay'anmey nu' QeD wIDelmeH mI'mey tlhatmeyvam DIlo'.

▣❯=≫◯☺

leko li sama e ijo nasa.

vay' jal nel meyrl'. jalwl'mey bIH meyrl'mey'e'.

⊣◯୫❯∸✕∸!

taso ijo musi li lon ala lon kin ?

'ach vay'mey Qatlh tu'lu"a' je ?

ᕈ✕≫ധ☱ᕈ ⊣ᕈധ☱:
ᆱL‿↟=❯∸✕∸

mi ante e wile sona mi. En mi wile sona:
selo pi linja nasin sama li lon ala lon ?

mu'tlheghmaj wItlhobbogh wIchoH 'ej maghel:
loS reD mey'mey Don tu'lu"a' ?

33

lon a !

HISlaH; blH tu'lu' !

selo pi linja nasin sama li lon.

loS reD mey'mey Don tu'lu'.

selo pi linja nasin sama li pini kulupu pi linja nasin ante tu.

cha' baSta' pIm yughbogh mI'QeD vIqraq bIH loS reD mey'mey Don'e'.

○ ⅍ :

ijo pana / ghantoH:

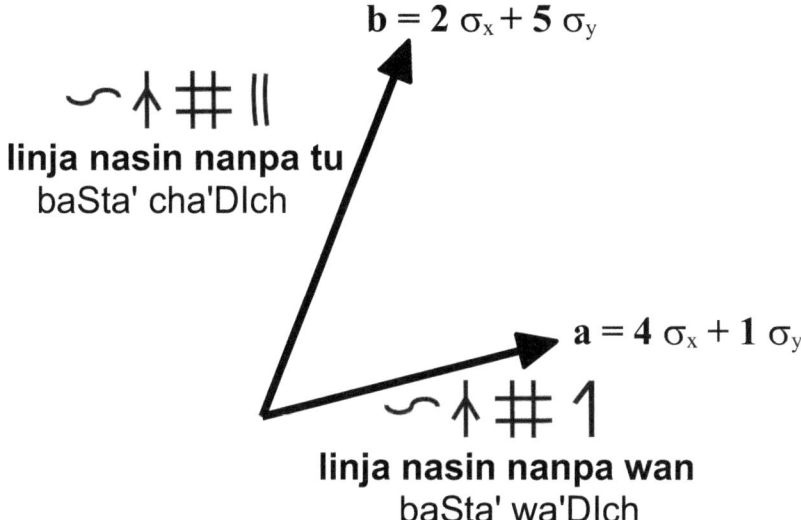

$$b = 2\,\sigma_x + 5\,\sigma_y$$

linja nasin nanpa tu
baSta' cha'DIch

$$a = 4\,\sigma_x + 1\,\sigma_y$$

linja nasin nanpa wan
baSta' wa'DIch

I ⁰⁰L ⌒↑↓‖ :

pini kulupu pi linja nasin ni tu:
cha' baSta'vam mI'QeD vIqraq :

$$\mathbf{a\,b} = (4\,\sigma_x + 1\,\sigma_y)\,(2\,\sigma_x + 5\,\sigma_y)$$

$$= 8\,\sigma_x^2 + 20\,\sigma_x\sigma_y + 2\,\sigma_y\sigma_x + 5\,\sigma_y^2$$

$$1 \qquad -\sigma_x\sigma_y \qquad 1$$

$$= 8 + 20\,\sigma_x\sigma_y - 2\,\sigma_x\sigma_y + 5$$

$$= 13 + 18\,\sigma_x\sigma_y$$

⌒↑#1>⸫≫⌒↑#‖)
↓>=≫ⅢL⌒↑=⁝

linja nasin nanpa wan li kulupu e linja nasin nanpa tu la ni li sama e selo pi linja nasin sama:

wa'logh baSta' cha'DIch boq'egh baSta' wa'DIch;
chen loS reD mey'vam Don:

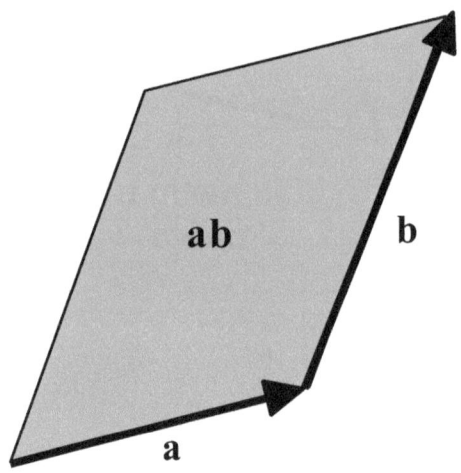

$$\mathbf{a\,b} = (4\,\sigma_x + 1\,\sigma_y)\,(2\,\sigma_x + 5\,\sigma_y) = 13 + 18\,\sigma_x\sigma_y$$

ⅢL⌒↑=>⅋≫#÷≫○⊚

selo pi linja nasin sama li jo e nanpa lon e ijo nasa.

ml' ruj, vay' jal je yugh loS reD mey' Don.

⊓⌐∟⌣⫪=⟩Ɛ⟫⊓℧

selo pi linja nasin sama li jo e sijelo musi.

Qur Qatlh ghaj loS reD mey' Don.

Ω∟ö⌗⟩ꝗ⟫⌑⊥

jan pi sona nanpa li pali e nimi sin.

pongmey chu' 'ogh mI'tejpu'.

⁒⌐∟Ⅰ⚬⚬⟩⌑⟫Ⅰ⚬⚬⊡

kipisi lon pi pini kulupu li nimi e pini kulupu insa.

'ay' ruj ghaj mI'QeD vIqraq.
'ay' rujvaD mI'QeD vIqraq qoD wIpong qoj
'ay' rujvaD qoD mI'QeD vIqraq wIpong.

$$\langle a\ b \rangle_0 = a \bullet b = 13$$

ᕒ⟩⊡⟫Ⅰ⚬⚬⊡⽮⊡◎

ona li sitelen e pini kulupu insa, kepeken sitelen sike.

mI'QeD vIqraq qoD wIghItlhmeH 'oDtu' rutlh wIlo'.

$$\text{I}\;\substack{\circ\\\circ\circ}\;\sqcup\;>\;\&\;\times\;\gg\;\uparrow\;\gg\;\uparrow\;\oplus$$

pini kulupu insa li jo ala e nasin e nasin ma.

roDSer lurgh joq ghajbe' mI'QeD vIqraq qoD.

ona li nanpa.

mI' 'oH.

$$\#\;\downarrow\;>\;\ddot{\text{w}}\;\gg\;\sqcup\;\cup\;\llcorner\;\frown\;\uparrow\;\|$$

nanpa ni li pana e insa lupa pi linja nasin tu.

cha' baSta' qubbID tajvaj Del mI'vam.

$$a^2 = (4\,\sigma_x + 1\,\sigma_y)^2 = 16 + 4\,\sigma_x\sigma_y - 4\,\sigma_x\sigma_y + 1 = 17$$

$$|a| = \sqrt{a^2} = 4.1231$$

$$a \bullet b = |a|\,|b|\cos\alpha = 13 \quad \Rightarrow \quad \cos\alpha = 0.5855$$
$$\alpha = 54.16°$$

$$|b| = \sqrt{b^2} = 5.3852$$

$$b^2 = (2\,\sigma_x + 5\,\sigma_y)^2 = 4 + 10\,\sigma_x\sigma_y - 10\,\sigma_x\sigma_y + 25 = 29$$

**jan pi sona nanpa pi pilin pona li pali
e nimi sin mute.**

pongmey chu' 'ogh mI'tejpu' Quch 'ej law' pongmey.

kipisi nasa pi pini kulupu li nimi e pini kulupu selo.

'ay' jal ghaj mI'QeD vIqraq.
'ay' jalvaD mI'QeD vIqraq Hur wIpong qoj
'ay' jalvaD Hur mI'QeD vIqraq wIpong.

$$\langle \mathbf{a}\,\mathbf{b} \rangle_2 = \mathbf{a} \wedge \mathbf{b} = \mathbf{18}\ \sigma_x \sigma_y$$

**ona li sitelen e pini kulupu selo, kepeken
sitelen nena.**

mI'QeD vIqraq Hur wIghItlhmeH ley' qIv wIlo'.

pini kulupu selo li jo e nasin tu anu nasin ma tu.

cha' roDSer, cha' lurgh joq ghaj mI'QeD vIqraq Hur.

ꝏ>‖⌣↑   ↓>‰⊚

ona li tu linja nasin.   **ni li kipisi nasa.**

cha'baSta' 'oH. 'ay' jal 'oHchu'.

‖⌣↑↓>ꙮ≫⊕ᴨ∟⌣↑=
ꝏ>⊕↑
+ꝏ>ꙮ≫∨⊕ᴨ∟⌣↑=

tu linja nasin ni li pana e ma selo pi linja nasin sama. ona li ma nasin. en ona li pana e suli ma selo pi linja nasin sama.

loS reD mey' Don qoD Del cha'baSta'vam.
lurghmey ghajbogh pa' qoD 'oH. 'ej menwl' Del 'oH.

I⊶ᴨ>=≫‰⊚
↓>=≫⊕ᴨ∟⌣↑=

pini kulupu selo = kipisi nasa
= ma selo pi linja nasin sama

ml'QeD vlqraq Hur = 'ay' jal
= lurghmey ghajbogh loS reD mey' Don qoD

$$A = a \wedge b = 18\ \sigma_x\sigma_y \quad \Rightarrow \quad |A| = \sqrt[4]{A^4} = 18$$

$$|A| = |a|\,|b|\sin\alpha = 4.1231 \cdot 5.3852 \cdot \sin 54.16° = 18$$

ᛒᚿ》⊔�𝖫☷#

mi kama e insa pi sona nanpa.

ml'QeD qolqoS wISIch. ml'QeD ghanroq wISIch.

θ◡#‖1꞉

lawa pona nanpa tu wan:

ml'QeD chut wejDIch:

$$\mathbf{a\,b} = \mathbf{a} \bullet \mathbf{b} + \mathbf{a} \wedge \mathbf{b}$$

☷𝖫☷#↓〉▭》Ɪ⁒𝖫θ◡

**toki pi sona nanpa ni li nimi e pini kipisi
pi lawa pona.**

witte'vamvaD ml'QeD wIrIvmeH chut tutlh wIpong.

¡◉》↓꞉

o lukin e ni / naDev yIlegh:

Garret Sobczyk: David Hestenes – The Early Years.
American Journal of Physics, Vol. 23, No. 10 (1993),
pp. 1290 – 1293.

ⵎ⌐⌣⥮=⟩=⟫#÷+◯ⓖ

selo pi linja nasin sama = nanpa lon + ijo nasa

loS reD mey' Don = ml' ruj + vay' jal

⇕

Ⓘ⸚⌐⌣⥮‖⟩=⟫Ⓘ⸚ᴗ+Ⓘ⸚ⵎ

**pini kulupu pi linja nasin tu
= pini kulupu insa + pini kulupu selo**

cha' baSta' ml'QeD vIqraq
= ml'QeD vIqraq qoD + ml'QeD vIqraq Hur

Ⓘ⸚⌐⌣⥮‖⟩Ɛ⟫Ⓘ⸚ᴗ⟫Ⓘ⸚ⵎ

**pini kulupu pi linja nasin tu li jo e pini kulupu insa
e pini kulupu selo.**

ml'QeD vIqraq qoD, ml'QeD vIqraq Hur je yugh
cha' baSta' ml'QeD vIqraq.

o lukin e ni / naDev yllegh:

David Hestenes: Oersted Medal Lecture 2002
– Reforming the Mathematical Language of Physics.
American Journal of Physics, Vol. 71, No. 2 (2003),
pp. 104 – 121.

ꙴ I :

sona pini / gher'ID ngaD:

$$a\, b = a \bullet b + a \wedge b$$

$$b\, a = b \bullet a + b \wedge a = a \bullet b - a \wedge b$$

ꙴ L ꙴ ♯ L ♯ ⚊⟩⫴ ≫ ꙴ L ꙴ ♯ L ♯ ‖⟩↓⟩ꙵ ≫ I °° ⊔•

toki pi sona nanpa pi nanpa wan li mute e toki pi sona nanpa pi nanpa tu la ni li pana e pini kulupu insa.

wItte' cha'DIch boq wItte' wa'DIch;
chen mI'QeD vIqraq qoD.

θ ∪ ♯ ‖ ‖ :

lawa pona nanpa tu tu:

mI'QeD chut loSDIch:

$$a \bullet b = \frac{1}{2}(a\, b + b\, a)$$

☸I┼:

sona pini namako / latlh gher'ID ngaD:

$$a\,b = \ a \bullet b + a \wedge b$$

$$-\,b\,a = -\,b \bullet a - b \wedge a = -\,a \bullet b + a \wedge b$$

☸L☸#L#1⟩ᵛ≫☸L☸# L#‖)↓⟩ѱ≫I⚬ᵒᵒ⊓⊓

toki pi sona nanpa pi nanpa wan li lili e toki pi sona nanpa pi nanpa tu la ni li pana e pini kulupu selo.

wItte' wa'DIch boqHa' wItte' cha'DIch; chen mI'QeD vIqraq Hur.

θ∪#∩:

lawa pona nanpa luka:

mI'QeD chut vaghDIch:

$$a \wedge b = \frac{1}{2}\left(a\,b - b\,a\right)$$

Ӫ└Ӫ#└⫽>⁒)ᏢᏗ꜔꜖≫Ⲓ⌣☉↓

**toki pi sona nanpa pi tu li lon la mi kepeken alasa
e pini pona, tenpo ni.**

DaH gher'ID wISamlaH cha' wItte' tu'lu'chugh.

¡꜖≫Ⲓ⌣

o alasa e pini pona !

gher'ID yISIm !

$$4\,x + 2\,y = 20$$
$$1\,x + 5\,y = 23 \qquad \Rightarrow \qquad x = ? \quad y = ?$$

¡°°≫⌢↑⌣

o kulupu e linja nasin pona !

'oplogh baSta'mey potlh boq'egh wItte'mey; chen
wItte'mey chu'. wItte'mey chu' tISam !

$$4\,x\,\sigma_x + 2\,y\,\sigma_x = 20\,\sigma_x$$
$$1\,x\,\sigma_y + 5\,y\,\sigma_y = 23\,\sigma_y$$

! ||| ≫↓   ! ∪ ≫ ᕋ

o mute e ni !   o pona e ona !

boq yISlm ! yInapmoH !

$$4 \, x \, \sigma_x + 2 \, y \, \sigma_x + 1 \, x \, \sigma_y + 5 \, y \, \sigma_y = 20 \, \sigma_x + 23 \, \sigma_y$$

$$\Rightarrow \quad 4 \, x \, \sigma_x + 1 \, x \, \sigma_y + 2 \, y \, \sigma_x + 5 \, y \, \sigma_y = 20 \, \sigma_x + 23 \, \sigma_y$$

$$\Rightarrow \quad (4 \, \sigma_x + 1 \, \sigma_y) \, x + (2 \, \sigma_x + 5 \, \sigma_y) \, y = 20 \, \sigma_x + 23 \, \sigma_y$$

! � ≫ ⌣ ↑

o alasa e linja nasin !

baSta'mey tISam !

$$\Rightarrow \quad (4 \, \sigma_x + 1 \, \sigma_y) \, x + (2 \, \sigma_x + 5 \, \sigma_y) \, y = 20 \, \sigma_x + 23 \, \sigma_y$$

$$a = 4 \, \sigma_x + 1 \, \sigma_y \qquad\qquad r = 20 \, \sigma_x + 23 \, \sigma_y$$

$$b = 2 \, \sigma_x + 5 \, \sigma_y$$

! ꓷ ≫ I ⧉ ⊓⊓ ∞

o alasa e pini kulupu selo ali !

Hoch mI'QeD vIqraq Hur yISam !

$$a\,b = (4\,\sigma_x + 1\,\sigma_y)\,(2\,\sigma_x + 5\,\sigma_y) = 13 + 18\,\sigma_x\sigma_y$$
$$b\,a = (2\,\sigma_x + 5\,\sigma_y)\,(4\,\sigma_x + 1\,\sigma_y) = 13 - 18\,\sigma_x\sigma_y$$
$$\Rightarrow \qquad\qquad a \wedge b = 18\,\sigma_x\sigma_y$$

$$a\,r = (4\,\sigma_x + 1\,\sigma_y)\,(20\,\sigma_x + 23\,\sigma_y) = 103 + 72\,\sigma_x\sigma_y$$
$$r\,a = (20\,\sigma_x + 23\,\sigma_y)\,(4\,\sigma_x + 1\,\sigma_y) = 103 - 72\,\sigma_x\sigma_y$$
$$\Rightarrow \qquad\qquad a \wedge r = 72\,\sigma_x\sigma_y$$

$$r\,b = (20\,\sigma_x + 23\,\sigma_y)\,(2\,\sigma_x + 5\,\sigma_y) = 103 + 54\,\sigma_x\sigma_y$$
$$b\,r = (2\,\sigma_x + 5\,\sigma_y)\,(20\,\sigma_x + 23\,\sigma_y) = 103 - 54\,\sigma_x\sigma_y$$
$$\Rightarrow \qquad\qquad r \wedge b = 54\,\sigma_x\sigma_y$$

!⊙»□∟Ⱋ ⟨Kↄ–'–ↄ⟩ↄↂↄ↑

!Ɛ»↑↶

o lukin e lipu pi jan Kasaman ! o jo e nasin ona !

loDnll paq yllaD ! Ho'DoSDaj yllo' !

Hermann Grassmann: Die Wissenschaft der extensiven Grösse oder die Ausdehnungslehre, eine neue mathematische Disciplin. Erster Theil, die lineale Ausdehnungslehre enthaltend. Verlag von Otto Wigand, Leipzig 1844, S. 70 – 73, § 45: Anwendung auf die Lösung algebr. Gleichungen, § 46: Lösung algebraischer Gleichungen.

ÖLÖ#LꝎ ⌐Kↄ–ⁱ–ↄↃↄ↑⌐

toki pi sona nanpa pi jan Kasaman:

loDnIl wItte'mey:

$$(a \wedge b)\, x = r \wedge b$$

$$\Rightarrow \qquad x = \frac{r \wedge b}{a \wedge b} = \frac{54\ \sigma_x\sigma_y}{18\ \sigma_x\sigma_y} = 3$$

$$(a \wedge b)\, y = a \wedge r$$

$$\Rightarrow \qquad y = \frac{a \wedge r}{a \wedge b} = \frac{72\ \sigma_x\sigma_y}{18\ \sigma_x\sigma_y} = 4$$

⋀⋀⋀ LÖ#>Ö∶
↓>∸Χ   ⌐>ΚΧ℅»⌒↑
+⌐>ΚΧ℅»‖⌒↑

monsuta pi sona nanpa li toki:
ni li lon ala. ona li ken ala kipisi e linja nasin.
en ona li ken ala kipisi e tu linja nasin.

jatlh mI'QeD veqlargh: mItbe' wItte'vam 'ej qarbe'.
'oplogh baSta' boqHa''eghlaHbe' vay'.
'ej 'oplogh cha'baSta' boqHa''eghlaHbe' vay'.

P⊙≫I∪

mi lukin e pini pona.

gher'IDmeymaj DI'ol.

$$4x + 2y = 20$$
$$1x + 5y = 23$$

$$\Rightarrow$$

$$4 \cdot 3 + 2 \cdot 4 = 20$$
$$1 \cdot 3 + 5 \cdot 4 = 23$$

$$x = 3 \qquad y = 4 \qquad \Rightarrow \qquad ↓>\,\dot{-}\quad \text{ni li lon.}$$

mlt gher'IDmeyvam 'ej qar.

sona nanpa pi pilin pona o !
mi ken kipisi e linja nasin
anu tu linja nasin.

'o ml'QeD Quch !
'oplogh baSta' boqHa''eghlaH vay'.
'ej 'oplogh cha'baSta' boqHa''eghlaH vay'.

monsuta Sinan li kama.
ona li utala e monsuta pi sona nanpa.

chol tlhIngan veqlargh. mI'QeD veqlargh Suv.

monsuta Sinan li toki:
linja nasin pona mi li jo ala e nasin taso.
mi olin e linja nasin ante.

jatlh tlhIngan veqlargh:
lurghmey leD ghajbe' baSta'meywIj potlh.
baSta'mey pIm vIparHa'.

o lukin e lipu ni !
o lukin e kipisi nanpa luka luka tu tu !

paqvam yIlegh ! 'ay' wa'maH loS yIlaD !

Martin Erik Horn: Das Klingonische Rechenbuch.
Teil 1: Algebra. Ein kleines Buch der Zahlen.
'ay' wa'DIch: ghIq ghIqtu' tu'. mI'mey paq mach.
BoD, Norderstedt 2024 (ISBN: 978-3-7583-6887-5).

monsuta Sinan li olin e linja nasin Sinan a !

tlhIngan baSta'mey muSHa'qu' tlhIngan veqlargh !

linja nasin Ewe $= -21\,\sigma_x + 25\,\sigma_y$
'ev baSta'

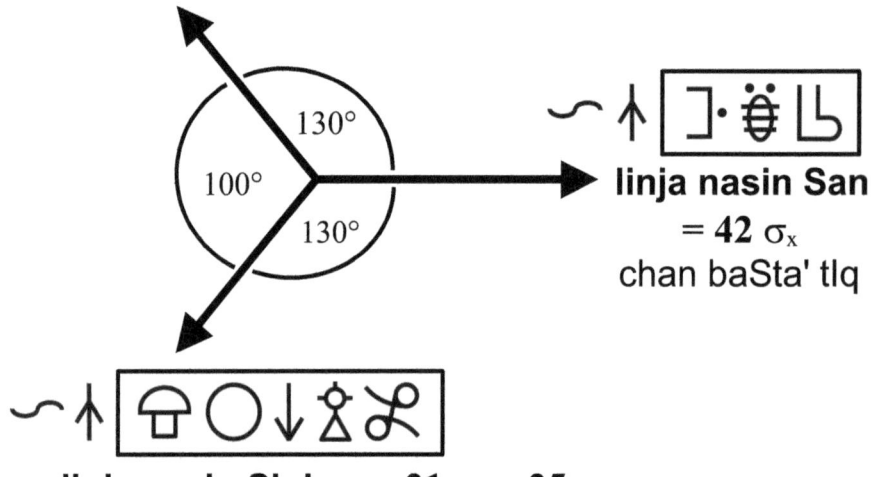

linja nasin San
$= 42\,\sigma_x$
chan baSta' tlq

linja nasin Sinje $= -21\,\sigma_x - 25\,\sigma_y$
tIng baSta'

insa lupa Sinan / tlhIngan tajvaj:

$$\cos\alpha = \frac{21}{\sqrt{21^2+25^2}} = 0.6432 \quad \Rightarrow \quad \alpha = 49.97° \approx 50°$$

$$\Rightarrow \quad \beta = 180° - \alpha \approx 130°$$

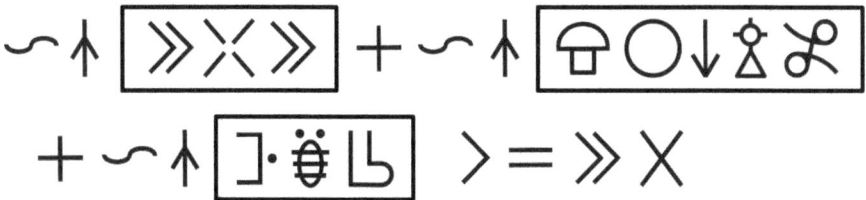

linja nasin Ewe + linja nasin Sinje + linja nasin San $= 0$

'ev baSta' + tlng baSta' + chan baSta' tlq $= 0$

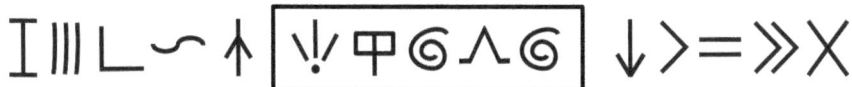

pini mute pi linja nasin Sinan ni li sama e ala.

nlb tlhlngan baSta'meyvam boq, pagh je.

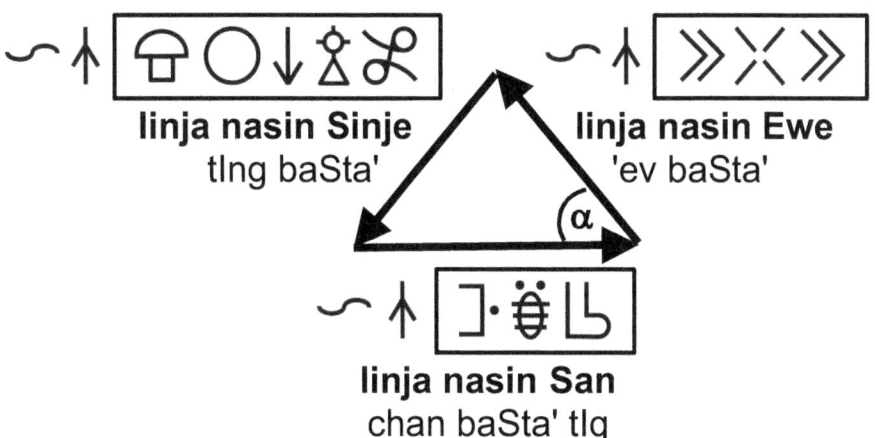

linja nasin Sinje
tlng baSta'

linja nasin Ewe
'ev baSta'

linja nasin San
chan baSta' tlq

⅄ |\!/ 平 ⑥ Λ ⑥| :

pali Sinan / tlhIngan Qu':

$$4\,x + 2\,y = 20$$
$$1\,x + 5\,y = 23 \qquad \Rightarrow \qquad x = ? \quad y = ?$$

o alasa e pini pona !
o kepeken e linja nasin Sinan !

gher'ID yISIm ! tlhIngan baSta'mey tIlo' !

o kulupu e linja nasin Sinan !

'oplogh tlhIngan baSta'mey boq'egh wItte'mey; chen wItte'mey chu'. wItte'mey chu' tISam !

$$4\,x\,(-21\,\sigma_x + 25\,\sigma_y) + 2\,y\,(-21\,\sigma_x + 25\,\sigma_y)$$
$$= 20\,(-21\,\sigma_x + 25\,\sigma_y)$$
$$1\,x\,(-21\,\sigma_x - 25\,\sigma_y) + 5\,y\,(-21\,\sigma_x - 25\,\sigma_y)$$
$$= 23\,(-21\,\sigma_x - 25\,\sigma_y)$$

o pona e ni !

yI'IHmoH !

$$-84 \text{ x } \sigma_x + 100 \text{ x } \sigma_y - 42 \text{ y } \sigma_x + 50 \text{ y } \sigma_y$$
$$= -420 \, \sigma_x + 500 \, \sigma_y$$
$$-21 \text{ x } \sigma_x - 25 \text{ x } \sigma_y - 105 \text{ y } \sigma_x - 125 \text{ y } \sigma_y$$
$$= -483 \, \sigma_x - 575 \, \sigma_y$$

o mute e ni !   o pona e ona !

boq yISIm ! yInapmoH !

$$-84 \text{ x } \sigma_x - 21 \text{ x } \sigma_x + 100 \text{ x } \sigma_y - 25 \text{ x } \sigma_y$$
$$-42 \text{ y } \sigma_x - 105 \text{ y } \sigma_x + 50 \text{ y } \sigma_y - 125 \text{ y } \sigma_y$$
$$= -420 \, \sigma_x - 483 \, \sigma_x + 500 \, \sigma_y - 575 \, \sigma_y$$

$$\Rightarrow \; -105 \text{ x } \sigma_x + 75 \text{ x } \sigma_y - 147 \text{ y } \sigma_x - 75 \text{ y } \sigma_y$$
$$= -903 \, \sigma_x - 75 \, \sigma_y$$

o alasa e linja nasin !

baSta'mey tISam !

$$\Rightarrow \quad (-\,\mathbf{105}\,\sigma_x + \mathbf{75}\,\sigma_y)\,x + (-\,\mathbf{147}\,\sigma_x - \mathbf{75}\,\sigma_y)\,y$$

$$= -\,\mathbf{903}\,\sigma_x - \mathbf{75}\,\sigma_y$$

$$\mathbf{a} = -\,\mathbf{105}\,\sigma_x + \mathbf{75}\,\sigma_y \qquad \mathbf{r} = -\,\mathbf{903}\,\sigma_x - \mathbf{75}\,\sigma_y$$

$$\mathbf{b} = -\,\mathbf{147}\,\sigma_x - \mathbf{75}\,\sigma_y$$

o alasa e pini kulupu selo ali !

Hoch mI'QeD vIqraq Hur yISam !

$$\mathbf{a\,b} = (-\,\mathbf{105}\,\sigma_x + \mathbf{75}\,\sigma_y)\,(-\,\mathbf{147}\,\sigma_x - \mathbf{75}\,\sigma_y)$$
$$= \mathbf{9810} + \mathbf{18900}\,\sigma_x\sigma_y$$

$$\mathbf{b\,a} = (-\,\mathbf{147}\,\sigma_x - \mathbf{75}\,\sigma_y)\,(-\,\mathbf{105}\,\sigma_x + \mathbf{75}\,\sigma_y)$$
$$= \mathbf{9810} - \mathbf{18900}\,\sigma_x\sigma_y$$

$$\Rightarrow \qquad \mathbf{a} \wedge \mathbf{b} = \mathbf{18900}\,\sigma_x\sigma_y$$

$$\mathbf{a\,r} = (-\,\mathbf{105}\,\sigma_x + \mathbf{75}\,\sigma_y)\,(-\,\mathbf{903}\,\sigma_x - \mathbf{75}\,\sigma_y)$$
$$= \mathbf{89190} + \mathbf{75600}\,\sigma_x\sigma_y$$

$$\mathbf{r\,a} = (-\,\mathbf{903}\,\sigma_x - \mathbf{75}\,\sigma_y)\,(-\,\mathbf{105}\,\sigma_x + \mathbf{75}\,\sigma_y)$$
$$= \mathbf{89190} - \mathbf{75600}\,\sigma_x\sigma_y$$

$$\Rightarrow \qquad \mathbf{a} \wedge \mathbf{r} = \mathbf{75600}\,\sigma_x\sigma_y$$

$$r\,b = (-903\,\sigma_x - 75\,\sigma_y)\,(-147\,\sigma_x - 75\,\sigma_y)$$
$$= 138366 + 56700\,\sigma_x\sigma_y$$

$$b\,r = (-147\,\sigma_x - 75\,\sigma_y)\,(-903\,\sigma_x - 75\,\sigma_y)$$
$$= 138366 - 56700\,\sigma_x\sigma_y$$

$$\Rightarrow \qquad\qquad r \wedge b = 56700\,\sigma_x\sigma_y$$

o jo e nasin pi jan Kasaman !

loDnll Ho'DoS yllo' !

toki pi sona nanpa pi jan Kasaman:

loDnll wItte'mey:

$$(a \wedge b)\,x = r \wedge b$$

$$\Rightarrow \qquad x = \frac{r \wedge b}{a \wedge b} = \frac{56700\,\sigma_x\sigma_y}{18900\,\sigma_x\sigma_y} = 3$$

$$(a \wedge b)\,y = a \wedge r$$

$$\Rightarrow \qquad y = \frac{a \wedge r}{a \wedge b} = \frac{75600\,\sigma_x\sigma_y}{18900\,\sigma_x\sigma_y} = 4$$

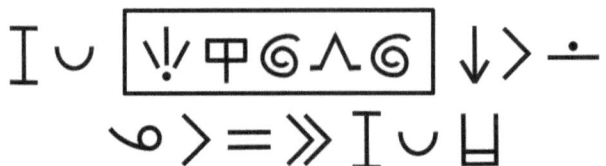

pini pona Sinan ni li lon.
ona li sama e pini pona open.

mIt tlhIngan gher'IDvam 'ej qar.
nIb tlhIngan gher'ID, gher'ID wa'DIch je.

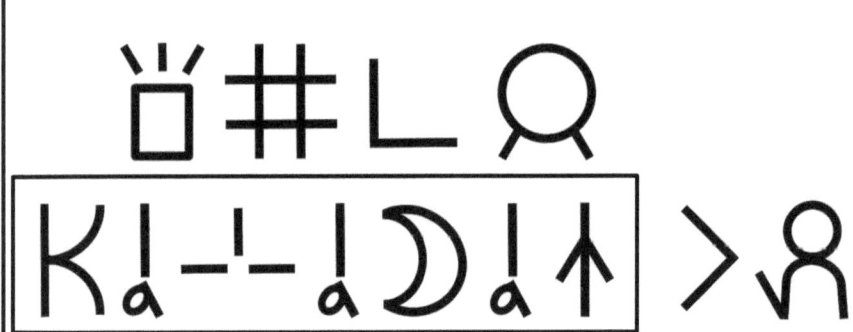

sona nanpa pi jan Kasaman li pali.

Qap loDnIl mI'QeD.

sona nanpa Sinan li pali.

Qap tlhIngan mI'QeD.

ni li selo pi linja nasin sama sin,
lon lipu sin, lon ni.

naDev (tenwal vebDaq) loS reD mey' Don tu'lu'.

↓〉＝✕≫ ⊓⊔∟ ⌣ ↑⊣

ni li sama ala e selo pi linja nasin taso.

naDev letbaQ tu'lu'be'.

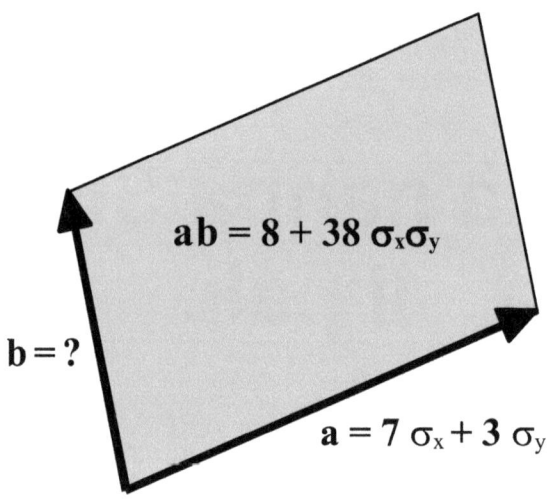

$ab = 8 + 38\ \sigma_x\sigma_y$

$b = ?$

$a = 7\ \sigma_x + 3\ \sigma_y$

⦙Ð⃗≫⌣↑#‖

o alasa e linja nasin nanpa tu !

baSta' cha'DIch tISam !

⊓∟⌒↑=¡°°≫⌒↑#1
¡⊔≫⌒↑#1

selo pi linja nasin sama o kulupu e linja nasin nanpa wan. o open e linja nasin nanpa wan !

wa'logh baSta' wa'DIch boq'eghjaj loS reD mey' Don.
wItte' bI'reSDaq baSta' wa'DIch tu'lu'jaj.

$$a\,a\,b = (7\,\sigma_x + 3\,\sigma_y)\,(8 + 38\,\sigma_x\sigma_y)$$

$$\sigma_x{}^2 = 1 \qquad\qquad -\sigma_x\sigma_y$$

$$= 56\,\sigma_x + 266\,\overbrace{\sigma_x\sigma_x}\sigma_y + 24\,\sigma_y + 114\,\overbrace{\sigma_y\sigma_x}\sigma_y$$

$$= 56\,\sigma_x + 266\,\sigma_y + 24\,\sigma_y - 114\,\underbrace{\sigma_x\sigma_y\sigma_y}$$

$$= 56\,\sigma_x + 290\,\sigma_y - 114\,\sigma_x \qquad \sigma_y{}^2 = 1$$

$$= -58\,\sigma_x + 290\,\sigma_y = a^2\,b$$

$$a^2 = (7\,\sigma_x + 3\,\sigma_y)\,(7\,\sigma_x + 3\,\sigma_y)$$

$$= 49 + 21\,\sigma_x\sigma_y - 21\,\sigma_x\sigma_y + 9$$

$$= 58$$

☼∟☼#>⁒≫#↓🕐↓

toki pi sona nanpa li kipisi e nanpa ni, tenpo ni.

DaH wa'logh mI'vam boqHa''egh wItte'.

$$\mathbf{b} = \frac{\mathbf{a}^2\,\mathbf{b}}{\mathbf{a}^2} = \frac{-\mathbf{58}\,\sigma_x + \mathbf{290}\,\sigma_y}{\mathbf{58}} = -\mathbf{1}\,\sigma_x + \mathbf{5}\,\sigma_y$$

jan pi sona nanpa li nimi e ni la ni li kipisi.

wItte'vamvaD 'ay'mey SubmaH lupong mI'tejpu'.

$$\mathbf{b} = \frac{\mathbf{a}}{\mathbf{a}^2}\,\mathbf{a}\,\mathbf{b} = \mathbf{a}^{-1}\,\mathbf{a}\,\mathbf{b}$$

$$\mathbf{a}^{-1} = \frac{\mathbf{a}}{\mathbf{a}^2} = \frac{\mathbf{7}}{\mathbf{58}}\,\sigma_x + \frac{\mathbf{3}}{\mathbf{58}}\,\sigma_y$$

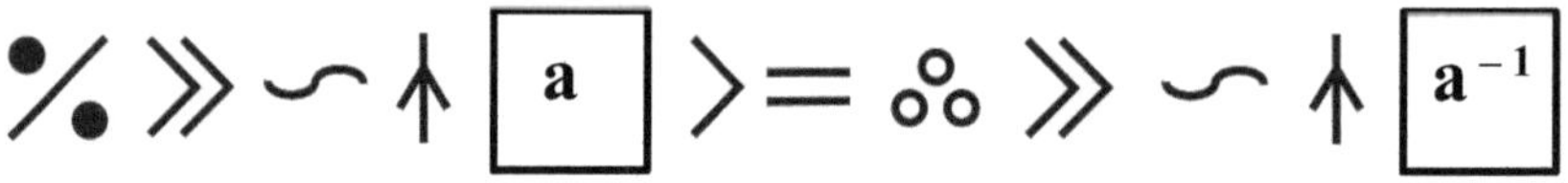

kipisi e linja nasin $\mathbf{a}$ = **kulupu e linja nasin** $\mathbf{a}^{-1}$

wa'logh baSta' $\mathbf{a}$ boqHa''egh vay'

= wa'logh baSta' $\mathbf{a}^{-1}$ boq'egh vay'

☿ ˅ ⁞

toki lili / qImlgh :

！⊔ ≫ ⁒

o open e kipisi !

wItte' bI'reSDaq 'ay'mey SubmaH tu'lu'jaj !

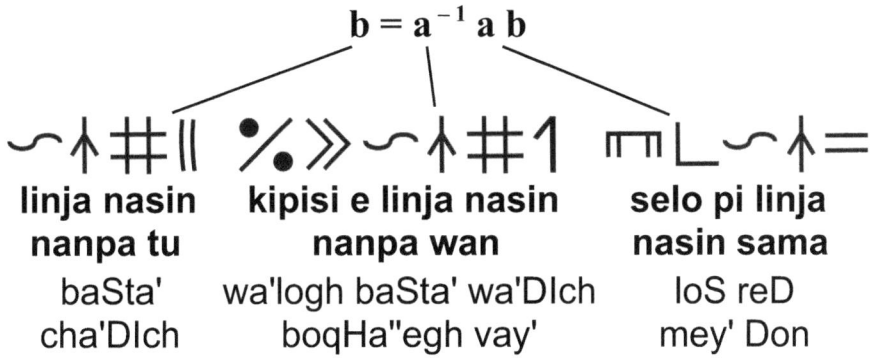

$$b = a^{-1}\, a\, b$$

linja nasin nanpa tu	kipisi e linja nasin nanpa wan	selo pi linja nasin sama
baSta' cha'DIch	wa'logh baSta' wa'DIch boqHa''egh vay'	loS reD mey' Don

！⊔ ≫ ⊓∟ ⌣ ↑ ＝

o open e selo pi linja nasin sama !

wItte' bI'reSDaq loS reD mey' Don tu'lu'jaj !

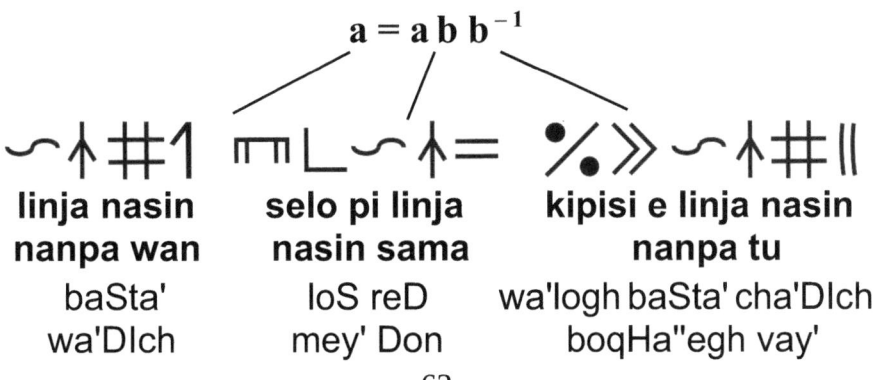

$$a = a\, b\, b^{-1}$$

linja nasin nanpa wan	selo pi linja nasin sama	kipisi e linja nasin nanpa tu
baSta' wa'DIch	loS reD mey' Don	wa'logh baSta' cha'DIch boqHa''egh vay'

$$\rho \odot \gg \mathrm{I} \cup$$

mi lukin e pini pona.

gher'IDmaj wl'ol.

$$a\,b = (7\,\sigma_x + 3\,\sigma_y)\,(-1\,\sigma_x + 5\,\sigma_y)$$

$$= -7\,\sigma_x^{\,2} + 35\,\sigma_x\sigma_y - 3\,\sigma_y\sigma_x + 15\,\sigma_y^{\,2}$$

$$= -7 + 35\,\sigma_x\sigma_y + 3\,\sigma_x\sigma_y + 15$$

$$= 8 + 38\,\sigma_x\sigma_y \quad \Rightarrow \quad \downarrow > \cdot \quad\quad \text{ni li lon.}$$

mlt gher'IDvam 'ej qar.

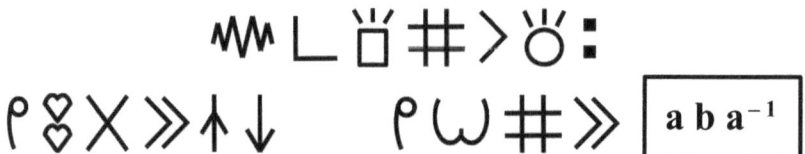

monsuta pi sona nanpa li toki:
mi olin ala e nasin ni. mi wile nanpa e $a\,b\,a^{-1}$.

jatlh ml'QeD veqlargh:

Dotlh cho'vam vlpar. $a\,b\,a^{-1}$ vlSlmqang.

ᴟ L ᵔ # ⟩

monsuta pi sona nanpa li nanpa.

jISlm ml'QeD veqlargh.

$$a\,b\,a^{-1} = \frac{1}{58}(8 + 38\,\sigma_x\sigma_y)\,(7\,\sigma_x + 3\,\sigma_y)$$

$$= \frac{1}{29}(4 + 19\,\sigma_x\sigma_y)\,(7\,\sigma_x + 3\,\sigma_y)$$

$$= \frac{1}{29}(28\,\sigma_x + 12\,\sigma_y + 133\,\sigma_x\sigma_y\sigma_x + 57\,\sigma_x\sigma_y\sigma_y)$$

$$= \frac{1}{29}(28\,\sigma_x + 12\,\sigma_y - 133\,\sigma_y + 57\,\sigma_x)$$

$$= \frac{1}{29}(85\,\sigma_x - 121\,\sigma_y)$$

$$= 2.93\,\sigma_x - 4.17\,\sigma_y$$

ᴟ L ᵔ # ! 6 ༃ ≫ ?

monsuta pi sona nanpa o, sina pali e seme ?

'o ml'QeD veqlargh, nuq Dachav ?

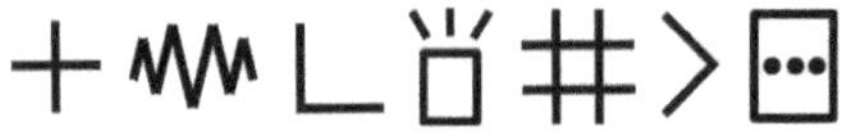

en monsuta pi sona nanpa li sitelen.

'ej wev mI'QeD veqlargh.

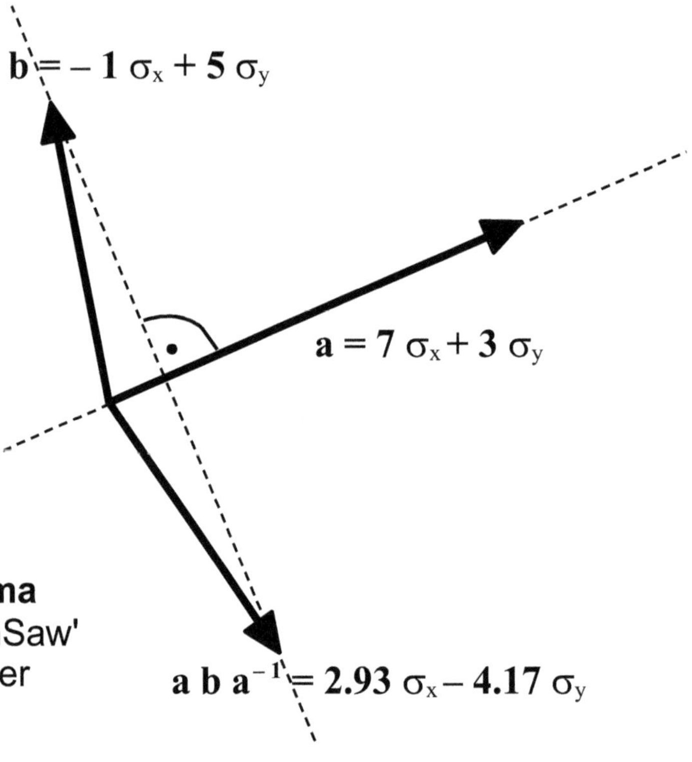

$\mathbf{b} = -\mathbf{1}\,\sigma_x + \mathbf{5}\,\sigma_y$
$\mathbf{a} = \mathbf{7}\,\sigma_x + \mathbf{3}\,\sigma_y$
$\mathbf{a\,b\,a^{-1}} = \mathbf{2.93}\,\sigma_x - \mathbf{4.17}\,\sigma_y$

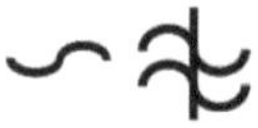

linja jasima
neSlo' tonSaw'
Qav rober

ni li jasima.

naDev neSlo' tonSaw'
Qav tu'lu'.

mi lukin e ni.

'oH wI'ol.

mi lukin e ni, kepeken nanpa tu wan.

wejlogh gher'IDvam wI'ol.

🕐1:

tenpo wan / He wa'DIch:

$$\vee \llcorner \curvearrowright \uparrow \nleftrightarrow > \omega = \gg \vee \llcorner \curvearrowright \uparrow \sqcup$$

suli pi linja nasin jasima li wile sama e suli pi linja nasin open.

nIbnIS baSta' narlu'chu'bogh lo'laHghach, baSta' wa'DIch lo'laHghach je.

$$b^2 = (-1\ \sigma_x + 5\ \sigma_y)\,(-1\ \sigma_x + 5\ \sigma_y)$$

$$= 1 + 25 = 26$$

$$b_{\text{ref}}^2 = \frac{1}{29^2}(85\ \sigma_x - 121\ \sigma_y)\,(85\ \sigma_x - 121\ \sigma_y)$$

$$= \frac{1}{841}(7225 + 14641)$$

$$= 26$$

$$\Rightarrow \quad \downarrow > \div$$

ni li lon.
mIt gher'IDvam 'ej qar.

⏱ ‖ :

tenpo tu / He cha'DIch:

ⵊⵊⵊⵊ∟⌣↑⚡+⌣↑⊔>ⵡ↑≫⌣⚡

**pini mute pi linja nasin jasima en linja nasin
open li wile nasin e linja jasima.**

baSta' narlu'chu'bogh boq baSta' wa'DIch; chen
baSta'mey boq. DonnIS baSta'mey boqvam, neSlo'
tonSaw' Qav rober je.

$$\mathbf{b}_{\text{ref}} + \mathbf{b} = \frac{1}{29}\,(85\,\sigma_x - 121\,\sigma_y) + (-1\,\sigma_x + 5\,\sigma_y)$$

$$= \frac{1}{29}\,(56\,\sigma_x + 24\,\sigma_y)$$

$$= \frac{8}{29}\,(7\,\sigma_x + 3\,\sigma_y)$$

$$= \frac{8}{29}\,\mathbf{a} \qquad \Rightarrow \qquad (\mathbf{b}_{\text{ref}} + \mathbf{b})\,\|\,\mathbf{a}$$

$$\Rightarrow \qquad \backsim\!\mathbf{o}>\mathbf{\&}\!\gg\!\uparrow =$$

ona li jo e nasin sama.
lurghmey nIb ghaj; Don.

$$\Rightarrow \qquad \downarrow > \div$$

ni li lon.
mIt gher'IDvam 'ej qar.

69

⊙∥1 :

tenpo tu wan / He wejDIch :

$$\text{I}^v \text{L} \smallfrown \text{↑} \text{⅋} + \smallfrown \text{↑} \text{⊔}$$
$$\text{>} \text{W} \text{⅋} \gg \text{↑⊢} \text{L} \smallfrown \text{⅋}$$

pini lili pi linja nasin jasima en linja nasin open li wile jo e nasin taso pi linja jasima.

baSta' narlu'chu'bogh boqHa' baSta' wa'DIch; chen baSta'mey chuq. leDnIS baSta'mey chuqvam, neSlo' tonSaw' Qav rober je.

$$\mathbf{b}_{ref} - \mathbf{b} = \frac{1}{29}\,(85\,\sigma_x - 121\,\sigma_y) - (-1\,\sigma_x + 5\,\sigma_y)$$

$$= \frac{1}{29}\,(114\,\sigma_x - 266\,\sigma_y)$$

$$\Rightarrow \quad \frac{1}{29}\,(114\,\sigma_x - 266\,\sigma_y) \bullet (7\,\sigma_x + 3\,\sigma_y) = 0$$

$$\Rightarrow \quad \cos\alpha = 0 \quad \Rightarrow \quad (\mathbf{b}_{ref} - \mathbf{b}) \perp \mathbf{a}$$

$$\Rightarrow \quad \smallfrown \text{>} \text{⅋} \gg \text{↑⊢}$$

ona li jo e nasin taso.
lurghmey leD ghaj.

$$\Rightarrow \quad \text{↓} \text{>} \dot{-}$$

ni li lon.
mIt gher'IDvam 'ej qar.

70

monsuta pi sona nanpa li jo ala e pilin pona.
ona li wile sona: ni li lon ala lon e jasima ?
ni li lon ala lon e jasima sike ?

QuchHa' mI'QeD veqlargh. tlhob:
naDev neSlo' tonSaw' Qav tu'lu''a' ? narta' baSta' ?
naDev jIrta'ghach tu'lu''a' ? jIrta''a' baSta' ?

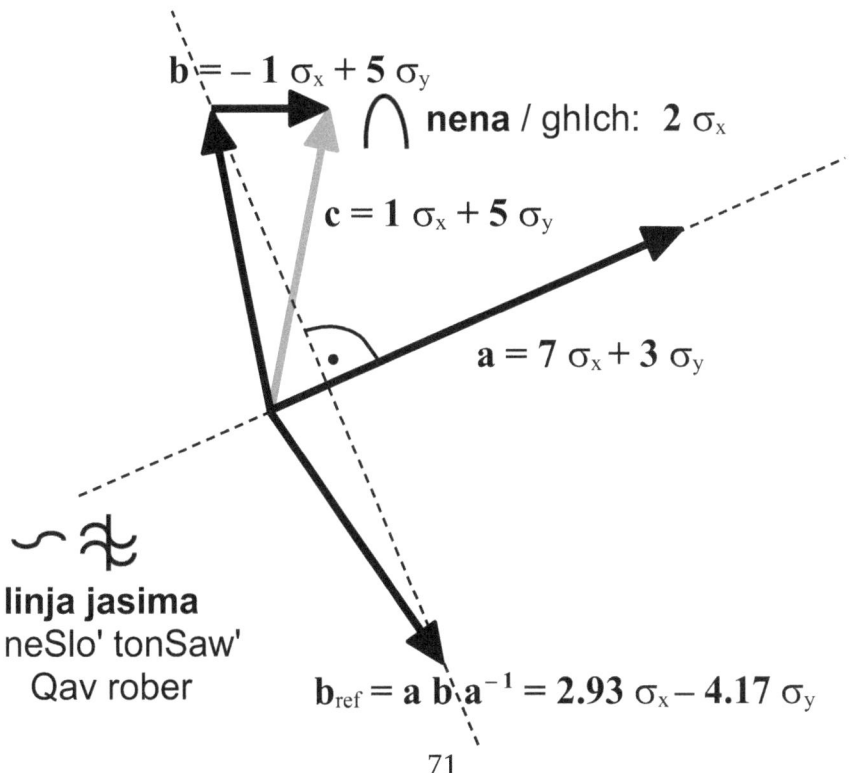

$$\mathbf{b} = -1\,\sigma_x + 5\,\sigma_y$$

nena / ghIch: $2\,\sigma_x$

$$\mathbf{c} = 1\,\sigma_x + 5\,\sigma_y$$

$$\mathbf{a} = 7\,\sigma_x + 3\,\sigma_y$$

linja jasima
neSlo' tonSaw'
Qav rober

$$\mathbf{b}_{\mathrm{ref}} = \mathbf{a}\,\mathbf{b}\,\mathbf{a}^{-1} = 2.93\,\sigma_x - 4.17\,\sigma_y$$

ᚹᛞᚷᚷᚷ

mi sona ala e ni.

wlSovbe'.

ᚷᚷᛞᚷᚷᚷ

tan ni la mi mute e nena.

vaj ghIch wIvev.

ᚷᚷᚷᚷᚷ c

mi jasima e linja nasin nena c.

c ghIch baSta' wInar.

$$a\,c\,a^{-1} = \frac{1}{58}(7\,\sigma_x + 3\,\sigma_y)(1\,\sigma_x + 5\,\sigma_y)(7\,\sigma_x + 3\,\sigma_y)$$

$$= \frac{1}{58}(22 + 32\,\sigma_x\sigma_y)(7\,\sigma_x + 3\,\sigma_y)$$

$$= \frac{1}{29}(11 + 16\,\sigma_x\sigma_y)(7\,\sigma_x + 3\,\sigma_y)$$

$$= \frac{1}{29}(77\,\sigma_x + 33\,\sigma_y + 112\,\sigma_x\sigma_y\sigma_x + 48\,\sigma_x\sigma_y\sigma_y)$$

$$= \frac{1}{29}(125\,\sigma_x - 79\,\sigma_y)$$

$$= 4.31\,\sigma_x - 2.72\,\sigma_y = c_{ref}$$

$$(125^2 + (-79)^2)/29^2 = 26 \quad \Rightarrow \quad$$ ᚷᚷᚷ **ni li lon.**

mIt gher'IDvam 'ej qar.

sitelen pini / mIllIogh naQ:

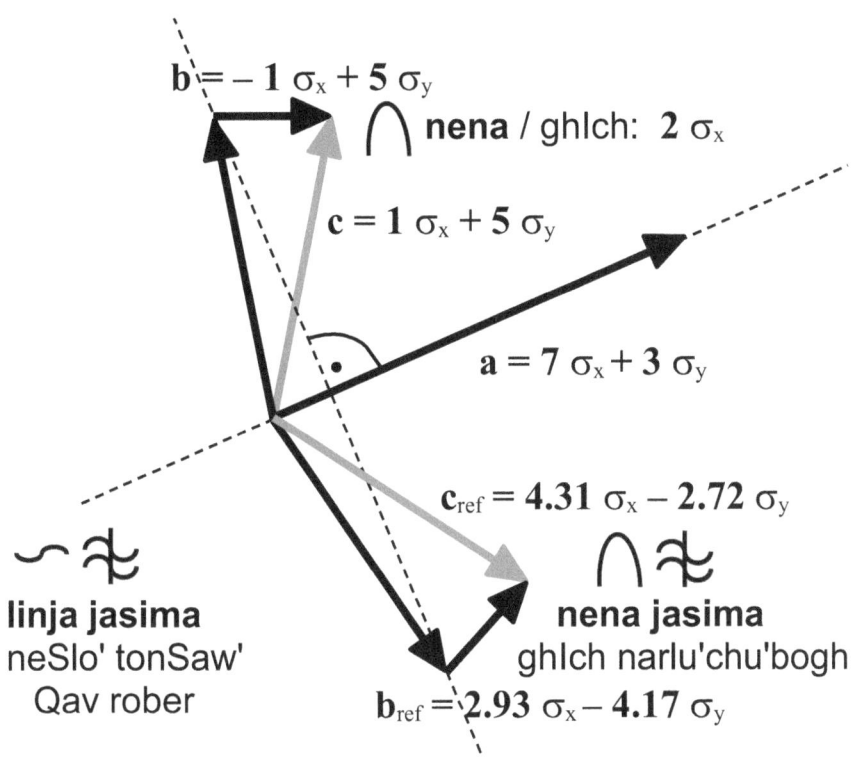

$b = -1\,\sigma_x + 5\,\sigma_y$

nena / ghIch: $2\,\sigma_x$

$c = 1\,\sigma_x + 5\,\sigma_y$

$a = 7\,\sigma_x + 3\,\sigma_y$

$c_{ref} = 4.31\,\sigma_x - 2.72\,\sigma_y$

linja jasima
neSlo' tonSaw'
Qav rober

nena jasima
ghIch narlu'chu'bogh

$b_{ref} = 2.93\,\sigma_x - 4.17\,\sigma_y$

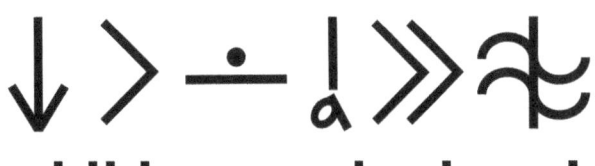

ni li lon a e jasima !

neSlo' tonSaw' Qav 'oHchu' !

θ∪#√∩1:

lawa póna nanpa luka wan:

ml'QeD chut javDIch:

$$a = a\,b\,b^{-1}$$

$$b = a^{-1}\,a\,b$$

θ∪#√∩‖:

lawa pona nanpa luka tu:

ml'QeD chut SochDIch:

$$a_{\text{ref}} = b\,a\,b^{-1} = b^{-1}\,a\,b$$

$$b_{\text{ref}} = a\,b\,a^{-1} = a^{-1}\,b\,a$$

jasima sike li sama lukin e seme kin ?

'ach nuq rur jIrta'ghach ?

monsuta pi sona nanpa li toki e ni tawa mi.

lutvam nujatlh mI'QeD veqlargh.

jasima sike li sama e jasima tu.

nIb wa' jIrta'ghach, cha' narta'ghach je.

**monsuta pi sona nanpa
li pona !**

lugh mI'QeD veqlargh !

☡#1↲∽☡ b :

jasima nanpa wan tawa linja jasima b:

b neSlo' tonSaw' Qav roberDaq neSlo' tonSaw' Qav wa'DIch:

$$\overbrace{b_{ref} = b\,b\,b^{-1}}^{1} = b = -1\,\sigma_x + 5\,\sigma_y$$

$$c_{ref} = b\,c\,b^{-1}$$

$$= \frac{1}{26}(-1\,\sigma_x + 5\,\sigma_y)(1\,\sigma_x + 5\,\sigma_y)(-1\,\sigma_x + 5\,\sigma_y)$$

$$= \frac{1}{26}(24 - 10\,\sigma_x\sigma_y)(-1\,\sigma_x + 5\,\sigma_y)$$

$$= \frac{1}{13}(12 - 5\,\sigma_x\sigma_y)(-1\,\sigma_x + 5\,\sigma_y)$$

$$= \frac{1}{13}(-12\,\sigma_x + 60\,\sigma_y + 5\,\sigma_x\sigma_y\sigma_x - 25\,\sigma_x\sigma_y\sigma_y)$$

$$= \frac{1}{13}(-37\,\sigma_x + 55\,\sigma_y)$$

$$= -2.85\,\sigma_x + 4.23\,\sigma_y$$

$$((-37)^2 + 55^2) / 13^2 = 26 \quad \Rightarrow \quad ↓>\,∴ \quad \text{ni li lon.}$$

mIt gher'IDvam 'ej qar.

ꙮ # ‖ ⌇ ⌒ ꙮ [a] :

jasima nanpa tu tawa linja jasima a:

a neSlo' tonSaw' Qav roberDaq neSlo' tonSaw' Qav cha'DIch:

$$\mathbf{b}_{rot} = \mathbf{a}\,\mathbf{b}_{ref}\,\mathbf{a}^{-1} = \mathbf{a}\,\mathbf{b}\,\mathbf{a}^{-1}$$

$$= \frac{1}{58}\,(7\,\sigma_x + 3\,\sigma_y)\,(-1\,\sigma_x + 5\,\sigma_y)\,(7\,\sigma_x + 3\,\sigma_y)$$

$$= \frac{1}{58}\,(8 + 38\,\sigma_x\sigma_y)\,(7\,\sigma_x + 3\,\sigma_y)$$

$$= \frac{1}{29}\,(4 + 19\,\sigma_x\sigma_y)\,(7\,\sigma_x + 3\,\sigma_y)$$

$$= \frac{1}{29}\,(85\,\sigma_x - 121\,\sigma_y)$$

$$= 2.93\,\sigma_x - 4.17\,\sigma_y$$

$$(85^2 + (-121)^2)\,/\,29^2 = 26 \quad \Rightarrow \quad \downarrow \rangle \div \qquad \text{ni li lon.}$$

mIt gher'IDvam 'ej qar.

$$\mathbf{c}_{rot} = \mathbf{a}\,\mathbf{c}_{ref}\,\mathbf{a}^{-1}$$

$$= \frac{1}{58\cdot 13}\,(7\,\sigma_x + 3\,\sigma_y)\,(-37\,\sigma_x + 55\,\sigma_y)\,(7\,\sigma_x + 3\,\sigma_y)$$

$$= \frac{1}{754}\,(-94 + 496\,\sigma_x\sigma_y)\,(7\,\sigma_x + 3\,\sigma_y)$$

$$= \frac{1}{377}\,(-47 + 248\,\sigma_x\sigma_y)\,(7\,\sigma_x + 3\,\sigma_y)$$

$$c_{rot} = \frac{1}{377} (415\ \sigma_x - 1877\ \sigma_y)$$

$$= 1.10\ \sigma_x - 4.98\ \sigma_y$$

$(415^2 + (-1877)^2) / 377^2 = 26 \Rightarrow$ ↓ > ∸ **ni li lon.**

mlt gher'IDvam 'ej qar.

⊡Ɪ :

sitelen pini / mlllogh naQ:

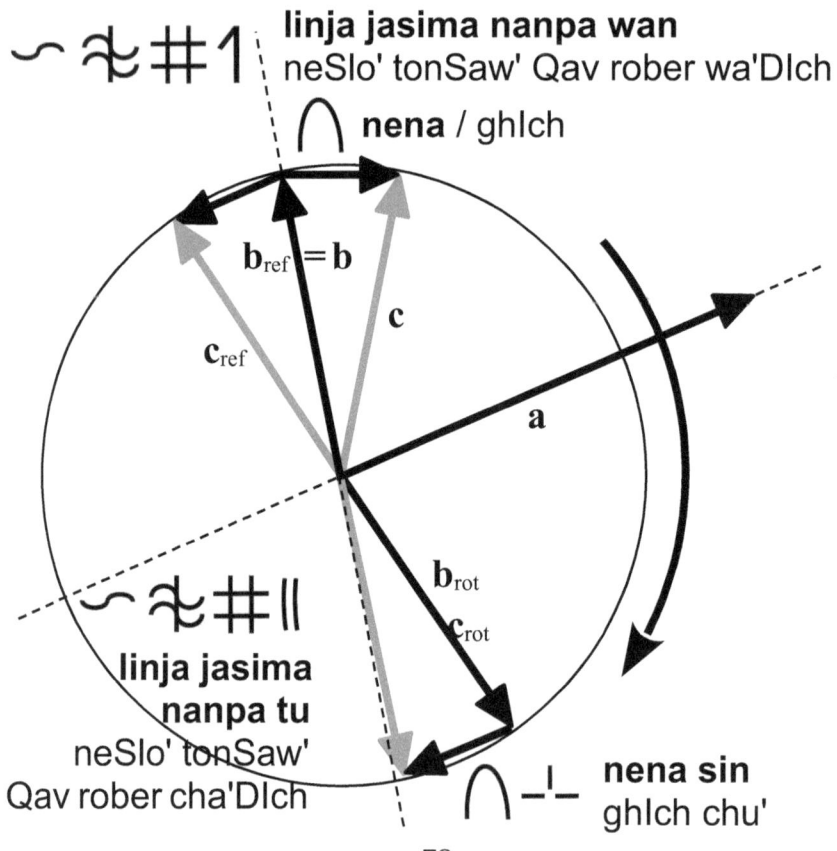

linja jasima nanpa wan
neSlo' tonSaw' Qav rober wa'DIch

nena / ghIch

$\mathbf{b}_{ref} = \mathbf{b}$

$\mathbf{c}$

$\mathbf{c}_{ref}$

$\mathbf{a}$

$\mathbf{b}_{rot}$

$\mathbf{c}_{rot}$

linja jasima
nanpa tu
neSlo' tonSaw'
Qav rober cha'DIch

nena sin
ghIch chu'

↓ ⟩ ⋅ ↓ₐ ≫ ⚡ ◎

ni li lon a e jasima sike !

jIrta'ghach 'oHchu' !

θ ⌣ ♯ √∩ ‖ 1 :

lawa pona nanpa luka tu wan:

mI'QeD chut chorghDIch:

$$c_{rot} = b\ c_{ref}\ b^{-1}$$
$$= b\ a\ c\ a^{-1} b^{-1}$$

↑ ✕ : **nasin ante** / lurgh Dop:

$$c_{rot*} = a\ c_{ref*}\ a^{-1}$$
$$= a\ b\ c\ b^{-1} a^{-1}$$

79

⌑∪∟♇◎❭𝟠≫∨↓ :

insa lupa pi jasima sike li jo e suli ni:

baSta' wa'DIch, baSta' jIrlu'chu'bogh je qubbID
tajvaj wIjuv:

$$a^2 = (7\,\sigma_x + 3\,\sigma_y)^2 = 49 + 21\,\sigma_x\sigma_y - 21\,\sigma_x\sigma_y + 9 = 58$$

$$|a| = \sqrt{a^2} = 7.6158$$

$$a \bullet b = |a|\,|b|\cos\alpha = 8 \quad \Rightarrow \quad \cos\alpha = 0.2060$$

$$\alpha = 78.11°$$

$$|b| = \sqrt{b^2} = 5.0990$$

$$b^2 = (-1\,\sigma_x + 5\,\sigma_y)^2 = 1 - 5\,\sigma_x\sigma_y + 5\,\sigma_x\sigma_y + 25 = 26$$

$$\beta_{rot} = 2\,\alpha = 2 \cdot 78.11° = 156.22° \quad \Rightarrow \quad \beta_{rot*} = -156.22°$$

⌑∪❭𝟠≫↑∟⾕◷)
⌑∪↓❭∹≫♯♇

**insa lupa li jo e nasin pi ilo tenpo la insa lupa ni
li lon e nanpa jasima.**

'Ingchugh tajvaj, Dop tajvajvam 'ej taH.

Ⲣⵙⵣⵉⵞⵐⵥⵔ#ⵏⵣⵙ

**mi lukin e pini pona, kepeken lipu nanpa
pi jasima sike.**

jIrta'ghach mI'mey tlhat wIlo'qangtaHvIS gher'ID wI'ol.

$$\begin{pmatrix} x_{rot} \\ y_{rot} \end{pmatrix} = \begin{pmatrix} \cos\beta & -\sin\beta \\ \sin\beta & \cos\beta \end{pmatrix} \begin{pmatrix} x \\ y \end{pmatrix}$$

nena sin ghIch chu'		-1 5
$\cos(-156.22°)$	$-\sin(-156.22°)$	2.93
$\sin(-156.22°)$	$\cos(-156.22°)$	-4.17

⇓

↓ ⟩ ⸱ **ni li lon.** $b_{rot} = 2.93\ \sigma_x - 4.17\ \sigma_y$
mIt gher'IDvam 'ej qar.

		1 5
$\cos(-156.22°)$	$-\sin(-156.22°)$	1.10
$\sin(-156.22°)$	$\cos(-156.22°)$	-4.98

⇓

↓ ⟩ ⸱ **ni li lon:** $c_{rot} = 1.10\ \sigma_x - 4.98\ \sigma_y$
mIt gher'IDvam 'ej qar.

**mi lukin e sewi, tenpo ni. mi lukin e sewi laso.
mi lukin e nasin ma sin.**

DaH DungDaq malegh. chalDaq malegh.
roDSer chu'Daq malegh.

nasin sin ni li nasin ma nanpa tu wan.

roDSer wejDIch 'oH lurghvam chu"e'.

**nasin ma nanpa tu wan li open e oko mi.
en ona li open e lawa mi.**

mInDu'maj poSmoH roDSer wejDIch.
'ej yabmaj poSmoH.

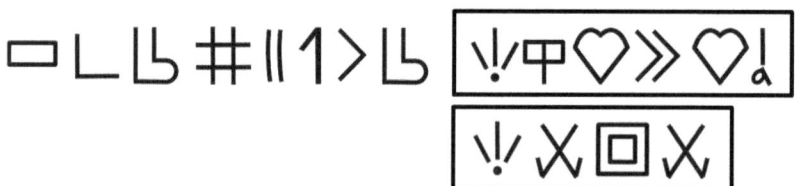

nimi pi noka nanpa tu wan li noka Sipepa Sulu.

baSta' potlh Dung 'oH qam chuq wejDIch pong'e'.

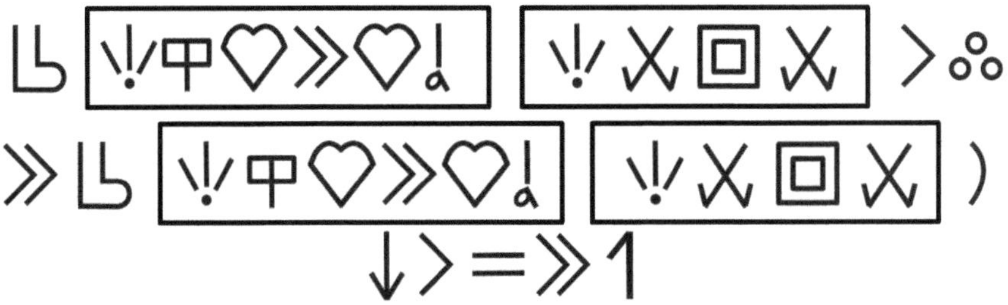

noka Sipepa Sulu li kulupu e noka Sipepa Sulu la ni li sama e wan.

wa'logh baSta' potlh Dung boq'egh wa' baSta' potlh Dung; chen wa'.

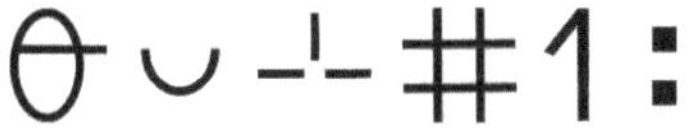

lawa pona sin nanpa wan:

mI'QeD chut Qa wa'DIch:

$$\sigma_x^2 = 1$$

$$\sigma_y^2 = 1$$

$$\sigma_z^2 = 1$$

ᚱꝋ » ↑⊕ ‖ 1 ☉ ↓

mi jo e nasin ma tu wan, tenpo ni.

DaH wej roDSer DIghaj.

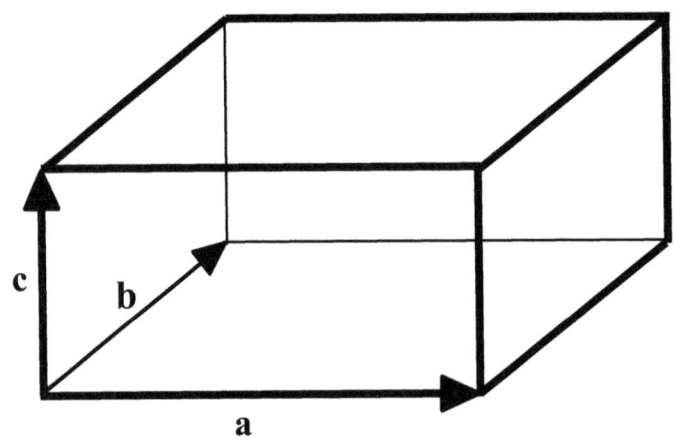

↓ > ⊓ L ⌒ ↑⊣

ni li sijelo pi linja nasin taso.

naDev letbaQ tut tu'lu'.

ᦉ > ꝋ » ⅃· ∩ 1
＋ ᦉ > ꝋ » ⌒ ∩ ∩ ‖

ona li jo e sinpin luka wan.
en ona li jo e linja luka luka tu.

jav reD ghaj 'ej wa'maH cha' HeH yamtaw ghaj.

jan Pisakojasa li lon a!

yIn peytlharghngongraS !

---

tenpo sin la mi alasa e jan Pisakojasa.

DaH peytlharghngongraS wInejqa'.

---

jan Pisakojasa li olin e leko.

meyrl'mey muSHa' peytlharghngongraS.

+Q 〔♡⌣‑'‑ₐK♡♡Gₐ‑'‑ₐ〕
〉♡♡》⌢ ⌣

en jan Pisakojasa li olin e linja insa.

'ej qoD yamtawmey muSHa' peytlharghngongraS.

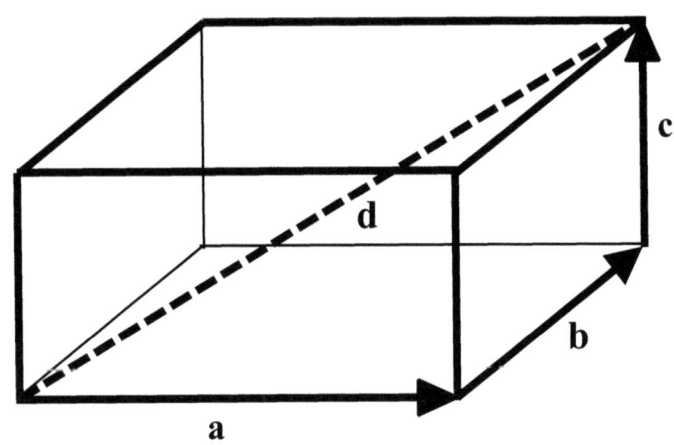

⌢↑⌣〉I ||| L⌢×||1

linja nasin insa li pini mute pi linja ante tu wan.

wej HeH yamtaw plm tu'lu'.

HeH yamtawmey boq 'oH qoD yamtaw baSta"e'.

$$a + b + c = d$$

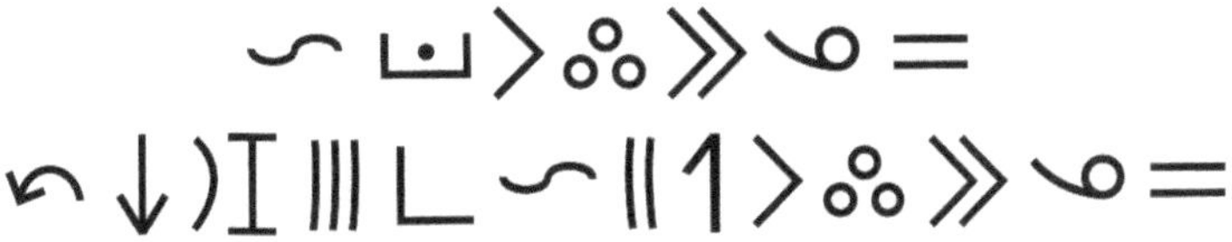

linja insa li kulupu e ona sama.
tan ni la pini mute pi linja tu wan li kulupu e ona sama.

wa'logh qoD yamtawvam boq'egh qoD yamtawvam. vaj wa'logh HeH yamtawmey boq boq'egh HeH yamtawmey boq.

$$\mathbf{d}^2 = (\mathbf{a} + \mathbf{b} + \mathbf{c})^2$$

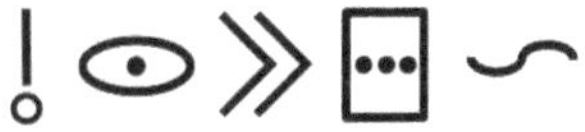

o lukin e sitelen linja !

naQjejHommey tIlegh !

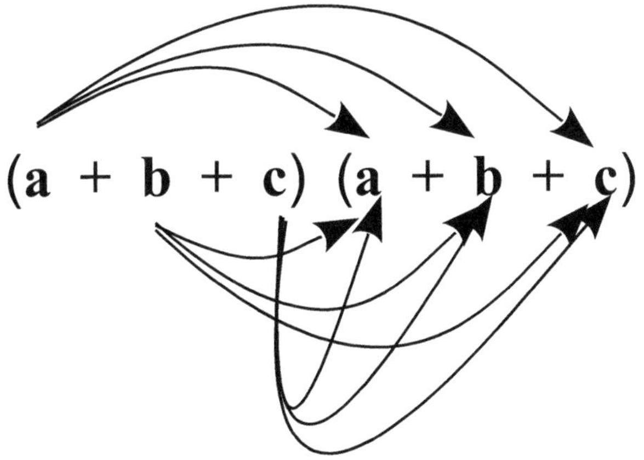

ᠹᠰ&≫‰∩‖‖

mi kama jo e kipisi luka tu tu.
Hut 'ay' DISuq.

‰ᠹ√∩‖‖‖>=≫

kipisi mi luka tu tu li sama e …

$$a\ a = a^2 \qquad a\ b \qquad a\ c$$
$$b\ a \qquad b\ b = b^2 \qquad b\ c$$
$$c\ a \qquad c\ b \qquad c\ c = c^2$$

… bIH Hut 'ay'meymaj'e'.

$$\Rightarrow \quad c^2 = (a + b + c)^2$$
$$= a^2 + a\ b + b\ a + b^2 + b\ c + c\ b + c^2 + c\ a + a\ c$$

**linja nasin ali li jo e nasin taso. tenpo sin la jan
Pisakojasa li kama e supa musi.**

leDchugh baSta'mey Hoch;
much yaH 'elqa' peytlharghngongraS.

⊙✕:

lukin ante / DIDay:

Q [♡⌣–ᵎ–↓K♡Ꙃ↓–ᵎ–↓] >ŏ:

jan Pisakojasa li toki:
jatlh peytlharghngongraS:

$$c^2 = a^2 \qquad\qquad + b^2 \qquad\qquad + c^2$$

$$c^2 = a^2 + \underbrace{a\,b + b\,a}_{0} + b^2 + \underbrace{b\,c + c\,b}_{0} + c^2 + \underbrace{c\,a + a\,c}_{0}$$

ŏ I:

sona pini / gher'ID ngaD:

$$\Rightarrow\quad a\,b + b\,a = 0 \qquad b\,c + c\,b = 0 \qquad c\,a + a\,c = 0$$

$$\Rightarrow\qquad a\,b = -\,b\,a \qquad b\,c = -\,c\,b \qquad c\,a = -\,a\,c$$

sitelen jasima
maQ Dop

ρ╳≫↑L⌒↑‖)⊡⚇>⅄

mi ante e nasin pi linja nasin tu la sitelen jasima li kama.

cha' baSta'mey Dotlh cho' wIyoymoHchugh, nargh maQ Dop.

ρ⊔⚇≫⌒↑◡

mi open kepeken e linja nasin pona.

DaH basta'mey potlh DIlo'choH.

$a = \sigma_x$

$b = \sigma_y$

$c = \sigma_z \quad \Rightarrow \quad$

$$a\,b = \sigma_x\,\sigma_y \qquad b\,c = \sigma_y\,\sigma_z \qquad c\,a = \sigma_z\,\sigma_x$$

$$b\,a = \sigma_y\,\sigma_x \qquad c\,b = \sigma_z\,\sigma_y \qquad a\,c = \sigma_x\,\sigma_z$$

ö I :

sona pini / gher'ID ngaD:

$$a\,b = -\,b\,a \qquad \Rightarrow \qquad \sigma_x\,\sigma_y = -\,\sigma_y\,\sigma_x$$

$$b\,c = -\,c\,b \qquad \Rightarrow \qquad \sigma_y\,\sigma_z = -\,\sigma_z\,\sigma_y$$

$$c\,a = -\,a\,c \qquad \Rightarrow \qquad \sigma_z\,\sigma_x = -\,\sigma_x\,\sigma_z$$

θ ∪ ⊣⊢ # ‖ :

lawa pona sin nanpa tu:

ml'QeD chut Qa cha'DIch:

$$\sigma_x\,\sigma_y = -\,\sigma_y\,\sigma_x$$

$$\sigma_y\,\sigma_z = -\,\sigma_z\,\sigma_y$$

$$\sigma_z\,\sigma_x = -\,\sigma_x\,\sigma_z$$

ρ & ≫ ⌣ ↑ ∪ ‖ 1 ⊙ ↓

mi jo e linja nasin pona tu wan, tenpo ni.
DaH wej baSta' potlh DIghaj.

θ ⌣ ✕ 〉 ⋀

lawa pona ante li awen.

Qabe' mI'QeD chutmey ratlh.

🕐 ⋀ 〉 ρ ⱳ ˙˙ö ≫ ∽ ∸ ⌊•⌋ ▯ ₋'₋

tenpo kama la mi wile toki e ona, lon insa lipu sin.

SIbI'Ha' latlh paq chu'Daq vIrIch.

🕐 ⋀ 〉 6 ↣ ≫ ⋁ ⊓ ⌐ ↟ ⊕ ‖ 1
∸ ⌊•⌋ ▯ ₋'₋ ↓

**tenpo kama la sina alasa e suli sijelo pi nasin
ma tu wan, lon insa lipu sin ni.**

paqvam chu'Daq wej roDSer ghajbogh yergh
lo'laHghach boSamlaH.

∽ 〉 ᘔ = ≫ ‖ 1 ⌣ ↟ ⌣ [I]

ona li jo sama e tu wan linja nasin pona I.

I wejbaSta' potlh nel 'oH.

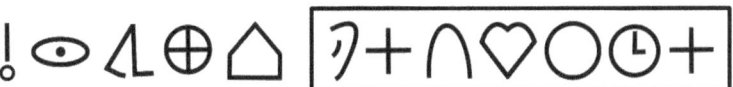

o lukin tawa ma tomo Kenpite !

qam blQtlq QI retlhDaq DuSaQ quv yIlegh !

"nanpa nasa li lon ala."

"Imaginary numbers are not real."

"mI'mey jal tu'lu'be'." "rujbe' mI'mey jal."

nanpa nasa li lon li lon ala.

mI'mey jal tu'lu'be'chu' 'ej mI'mey jal tu'lu'chu'.

tu wan linja nasin pona li lon:

wejbaSta' potlh tu'lu'chu':

$$I = \sigma_x \sigma_y \sigma_z$$

93

ᔆ ⟩ ◯ ⵚ

ona li ijo nasa.
vay' jal 'oH.

$$\mathbf{I} = \sigma_x \sigma_y \sigma_z \quad \Rightarrow \quad \mathbf{I}^2 = (\sigma_x \sigma_y \sigma_z)^2$$

$$= \sigma_x \sigma_y \sigma_z \, \sigma_x \sigma_y \sigma_z$$

$$= - \sigma_x \sigma_y \sigma_z \, \sigma_x \sigma_z \sigma_y$$

$$= + \sigma_x \sigma_y \sigma_z \sigma_z \sigma_x \sigma_y \qquad \sigma_z^2 = 1$$

$$= + \sigma_x \sigma_y \sigma_x \sigma_y$$

$$= - \sigma_x \sigma_x \sigma_y \sigma_y \qquad \sigma_y^2 = 1$$

$$= - \sigma_x \sigma_x \qquad \sigma_x^2 = 1$$

$$= - 1$$

"… we get … exactly the algebra of the Pauli spin matrices used in the quantum mechanics of spin-1/2 particles!"

ᏢᏗᏋ≫ᵒᵒ# 🄰♡ᶦ⟩ᶿ⟩⌣🄱 ∟
🄰#∟⚗◎⌣

mi kama jo e sona nanpa Paluli pi lipu nanpa
pi jasima sike insa.

qoD DIngta'ghach Delbogh mI'mey tlhatmey ghajbogh
paw'll' mI'QeD wISuq.

$$I\,\sigma_x = \sigma_x\,\sigma_y\,\sigma_z\,\sigma_x = -\,\sigma_x\,\sigma_y\,\sigma_x\,\sigma_z = \sigma_x\,\sigma_x\,\sigma_y\,\sigma_z = \sigma_y\,\sigma_z$$

$$I\,\sigma_y = \sigma_x\,\sigma_y\,\sigma_z\,\sigma_y = -\,\sigma_x\,\sigma_y\,\sigma_y\,\sigma_z = -\,\sigma_x\,\sigma_z = \sigma_z\,\sigma_x$$

$$I\,\sigma_z = \sigma_x\,\sigma_y\,\sigma_z\,\sigma_z = \sigma_x\,\sigma_y$$

θ◡#√∩‖ :

lawa pona nanpa luka tu tu:

mI'QeD chut HutDIch:

$$\sigma_x\,\sigma_y = I\,\sigma_z$$

$$\sigma_y\,\sigma_z = I\,\sigma_x$$

$$\sigma_z\,\sigma_x = I\,\sigma_y$$

○○L̃X̃>⊡≫↓℧□#

jan sona pi jaki ala li sitelen e ni, kepeken lipu nanpa.

mI'QeDvam ghItlhmeH mI'mey tlhatmey lo' tejpu' boch.

$$\sigma_x = \begin{pmatrix} 0 & 1 \\ 1 & 0 \end{pmatrix} \qquad \sigma_y = \begin{pmatrix} 0 & -i \\ i & 0 \end{pmatrix} \qquad \sigma_z = \begin{pmatrix} 1 & 0 \\ 0 & -1 \end{pmatrix}$$

			σ_y		σ_z	
$\sigma_x \sigma_y \sigma_z = ?$			0	−i	1	0
			i	0	0	−1
σ_x	0	1	i	0	i	0
	1	0	0	−i	0	i

$$\Rightarrow \qquad I = \begin{pmatrix} i & 0 \\ 0 & i \end{pmatrix}$$

$$\Rightarrow \qquad I^2 = \begin{pmatrix} -1 & 0 \\ 0 & -1 \end{pmatrix} = -1$$

$I^2 = ?$		i	0
		0	i
i	0	−1	0
0	i	0	−1

$$+\|\frown\uparrow\cup\rangle = \gg$$

en tu linja nasin pona li sama e …

'ej …

$$\sigma_z\sigma_x = \begin{pmatrix} 0 & 1 \\ -1 & 0 \end{pmatrix} = I\,\sigma_y = \begin{pmatrix} i & 0 \\ 0 & i \end{pmatrix}\begin{pmatrix} 0 & -i \\ i & 0 \end{pmatrix}$$

$$\sigma_y\sigma_z = \begin{pmatrix} 0 & i \\ i & 0 \end{pmatrix} = I\,\sigma_x$$

$$\begin{array}{c|c|c}
 & \sigma_x & \begin{matrix} 0 & 1 \\ 1 & 0 \end{matrix} \\
\hline
\sigma_z & \begin{matrix} 1 & 0 \\ 0 & -1 \end{matrix} & \begin{matrix} 0 & 1 \\ -1 & 0 \end{matrix} \\
\hline
\sigma_y & \begin{matrix} 0 & -i \\ i & 0 \end{matrix} & \begin{matrix} 0 & i \\ i & 0 \end{matrix} \\
\hline
\sigma_x & \begin{matrix} 0 & 1 \\ 1 & 0 \end{matrix} & \begin{matrix} i & 0 \\ 0 & -i \end{matrix}
\end{array}$$

$$I\,\sigma_x = \begin{pmatrix} i & 0 \\ 0 & i \end{pmatrix}\begin{pmatrix} 0 & 1 \\ 1 & 0 \end{pmatrix}$$

$$\sigma_x\sigma_y = \begin{pmatrix} i & 0 \\ 0 & -i \end{pmatrix} = I\,\sigma_z = \begin{pmatrix} i & 0 \\ 0 & i \end{pmatrix}\begin{pmatrix} 1 & 0 \\ 0 & -1 \end{pmatrix}$$

… bIH cha'baSta'mey'e'.

ᔆᖾ╫ ≫ ∞ ᖾ☐╫

mi kepeken nanpa e ale, kepeken lipu nanpa.

mI'mey tlhatmey DI'DIlo' Hoch wISImlaH.

○ᘏᔆ>ᗰ∟∿↑=

ijo pana mi li selo pi linja nasin sama.

loS reD mey' Don 'oH ghantoHmaj'e'.

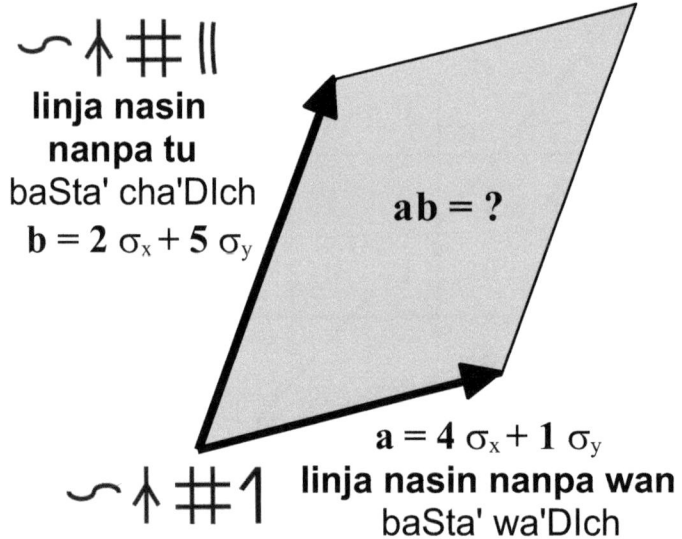

∿↑╫‖

**linja nasin
nanpa tu**
baSta' cha'DIch
$b = 2\,\sigma_x + 5\,\sigma_y$

$ab = ?$

$a = 4\,\sigma_x + 1\,\sigma_y$

∿↑╫1 **linja nasin nanpa wan**
baSta' wa'DIch

Ɪ∪∟–╵–Χ:

pini pona pi sin ala / gher'ID ngo':

$$a\,b = (4\,\sigma_x + 1\,\sigma_y)(2\,\sigma_x + 5\,\sigma_y) = 13 + 18\,\sigma_x\sigma_y$$

⊡ –ˈ– ⫶

sitelen sin / ghIqtu' chu':

$$\Rightarrow \quad a = 4\,\sigma_x + 1\,\sigma_y = 4 \begin{pmatrix} 0 & 1 \\ 1 & 0 \end{pmatrix} + 1 \begin{pmatrix} 0 & -i \\ i & 0 \end{pmatrix}$$

$$= \begin{pmatrix} 0 & 4-i \\ 4+i & 0 \end{pmatrix}$$

$$b = 2\,\sigma_x + 5\,\sigma_y = 2 \begin{pmatrix} 0 & 1 \\ 1 & 0 \end{pmatrix} + 5 \begin{pmatrix} 0 & -i \\ i & 0 \end{pmatrix}$$

$$= \begin{pmatrix} 0 & 2-5i \\ 2+5i & 0 \end{pmatrix}$$

Ⅰ ⸬ L ⌣ ↑ ‖ ⫶

pini kulupu pi linja nasin tu:
cha' baSta' mI'QeD vIqraq:

$\Rightarrow$ a b = ?	0	2 − 5 i
	2 + 5 i	0
0 4 − i	13 + 18 i	0
4 + i 0	0	13 − 18 i

I∪⌐ :

pini pona sin / gher'ID chu':

$$\mathbf{a\,b} = \begin{pmatrix} 13 + 18\,i & 0 \\ 0 & 13 - 18\,i \end{pmatrix}$$

$$= 13 \begin{pmatrix} 1 & 0 \\ 0 & 1 \end{pmatrix} + 18 \begin{pmatrix} i & 0 \\ 0 & -i \end{pmatrix}$$

$$= 13 + 18\ \sigma_x\sigma_y$$

$$\begin{pmatrix} 1 & 0 \\ 0 & 1 \end{pmatrix} = 1$$

¡)□#↓⟩=≫1

kin la lipu nanpa ni li sama e wan.
vabDot wa' 'oH ml'mey tlhatvam'e'.

I∪⌐⟩=≫I∪⊔

pini pona sin li sama e pini pona open.
nlb gher'ID chu', gher'ID ngo' je.

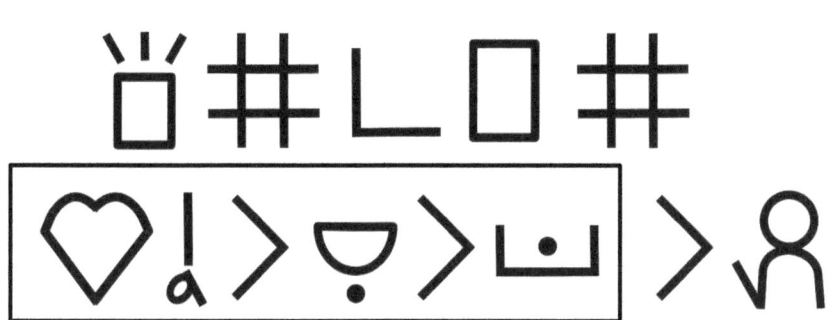

sona nanpa pi lipu nanpa Paluli li pali.

Qap mI'mey tlhatmey ghajbogh
paw'lI' mI'QeD'e'.

ni li pona lukin.
'IHqu'chu'.

**tan ni la mi awen e sona nanpa Kijaki.
ona li pona lukin kin.**

DI'raq mI'QeD wIloSmeH meqmaj 'oHchu'.
'IHqu'chu' je.

kama insa pi lipu sin:
ngaS paq chu':

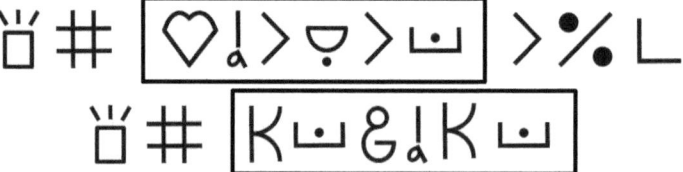

sona nanpa Kijaki li kipisi pi sona nanpa Paluli.

DI'raq mI'QeD 'ay' 'oH paw'lI' mI'QeD'e'.

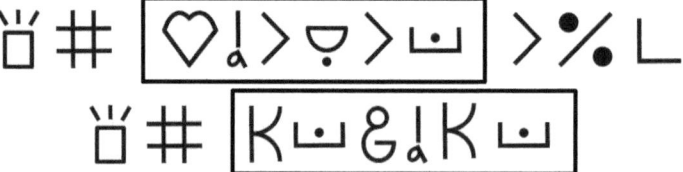

sona nanpa Paluli li kipisi pi sona nanpa Kijaki.

paw'lI' mI'QeD 'ay' 'oH DI'raq mI'QeD'e'.

kon nimi / mu'tay'

mi kepeken e nimi kama:
mu'mey vI'lo. naDev vI'agh:

ante
different, to change

choH / pIm
verschieden, ändern

ijo jasima
negative quantity

vay' Dop / Dopwl'
negative Größe

ijo lon
real quantity

vay' ruj /rujwl'
reelle Größe

ijo musi
complex quantity

vay' Qatlh / Qatlhwl'
komplexe Größe

ijo nasa
imaginary quantity

vay' jal / jalwl'
imaginäre Größe

ijo pana
example

ghantoH
Beispiel

insa lupa
the inside of the
orrifice, angle

tajvaj
das Innere der Öffnung,
Winkel

jan pi sona nanpa
mathematician

mI'tej
Mathematiker/in

jasima

Dop / narta'ghach
neSlo' tonSaw' Qav
(letzte Stellung eines kleinen Spiegels /
last stance of a small mirror)

opposite, minus,
negative, reflection

entgegengesetzt, minus
negativ, Reflexion

103

⚵◎	**jasima sike** rotation	jIrta'ghach Rotation, Drehung
⚵◎⊡	**jasima sike insa** interior rotation, spin	qoD DIngta'ghach innere Rotation, Spin
ℰ	**jo** to have, to possess	ghaj haben, besitzen
⋀ℰ	**kama jo** to get	Suq / Hev erhalten
⁒	**kipisi** part, term, to split to divide, division	'ay' / boqHa''egh Teil, Term, teilen dividieren, Division
⁒÷	**kipisi lon** real part, real term	'ay' ruj reeler Teil, reeller Term
⁒ⓖ	**kipisi nasa** imaginary part, imaginary term	'ay' jal imaginärer Teil, imaginärer Term
⚬⚬⚬	**kulupu** to form a group, to multiply, times, multiplication	boq'egh gruppieren, multiplizieren, mal, Multiplikation
θ	**lawa** rule	chut / ngangHa'wI' Regel
θ⌣	**lawa pona** basic rule	mI'QeD chut Grundregel
θ♀	**lawa tonsi** cha' 'ay'vaD mI'QeD chut binomial theorem Binomialtheorem
▣	**leko** square	meyrl' Quadrat

104

▣ ⏝	**leko pona** unit square, base square	**meyrI' potlh** Einheitsquadrat, Basisquadrat
∨	**lili** small, short, to de- crease, to subtract, subtraction	**mach / wII / qer QaHa' / boqHa'** klein, kurz, verringern, subtrahieren, Subtraktion
⌒	**linja** line, edge	**yamtaw / tlhegh / HeH yamtaw** Linie, Kante
⌒ ⊡	**linja insa** diagonal line	**qoD yamtaw** Diagonale
⌒ ⚐	**linja jasima** axis of reflection	**neSlo' tonSaw' Qav rober / narwI'** Reflexionsachse
⌒ ↑	**linja nasin** line with direction, vector	**baSta'** Linie mit Richtung, Vektor
⌒ ↑ ⊡	**linja nasin insa** vector of the diagonal line	**qoD baSta' / qoD yamtaw baSta'** Diagonalenvektor
⌒ ↑ ⚐	**linja nasin jasima** reflected vector	**baSta' narlu'chu'bogh** reflektierter Vektor
⌒ ↑ ‖‖	**linja nasin mute** several vectors	**baSta'mey puS** mehrere Vektoren

⌒↑◡ **linja nasin pona** baSta' potlh
unit vector, base Einheitsvektor,
vector Basisvektor

□# **lipu nanpa** ml'mey tlhat
matrix Matrix

□#∟↯◎ **lipu nanpa pi jasima sike**
jlrta'ghach ml'mey tlhat
matrix of rotation Rotationsmatrix

⊕ **ma** qoD / pa' qoD
area Fläche

⊕↑ **ma nasin** lurghmey ghajbogh pa' qoD
oriented area orientierte Fläche

||| **mute** puS / ghurmoH /
QamoH / boq
several, to increase mehrere, erhöhen
to add to, addition addieren zu, Addition

nanpa ml'
number Zahl

#↯ **nanpa jasima** ml' Dop
negative number negative Zahl

#⨪ **nanpa lon** ml' ruj
real number reelle Zahl

#ʊ **nanpa musi** ml' Qatlh
complex number komplexe Zahl

#ᓂ **nanpa nasa** ml' jal
imaginary number imaginäre Zahl

106

↑	**nasin** way, direction, orientation, order	taw / He / lurgh / cho' / Dotlh cho' Weg, Richtung, Orientierung, Reihenfolge
↑⚡⅃⊕◐	**nasin jasima pi ilo tenpo** Dotlh nam / letbaQ nam anti-clockwise orientation	Orientierung entgegen dem Uhrzeigersinn
↑⊕	**nasin ma** direction of space, dimension	roDSer Richtung der Welt, Dimension
↑⅃⊕◐	**nasin pi ilo tenpo** clockwise orientation	Dotlh 'Ing / letbaQ 'Ing Orientierung im Uhrzeigersinn
↑=	**nasin sama** parallel	Don parallel
↑⊣	**nasin taso** perpendicular, orthogonal	leD senkrecht, orthogonal
⅃	**noka** foot, step	qam / qam chuq Fuß, Schritt
⅃◁ᴛᴛ	**noka tawa supa** horizontal step, step into x-direction	qam chuq SaS horizontaler Schritt, Schritt in x-Richtung
⅃◁⅂·	**noka tawa sinpin** step forward	qam chuq tlhop Schritt nach vorne
⅃◁⌐	**noka tawa sewi** vertical step	qam chuq chong vertikaler Schritt

	pini kipisi end ef splitting, de-composition, quotient	ml'QeD lagh Ende der Teilung, Zerlegung, Quotient
I ⁒		

pini kipisi pi lawa pona

ml'QeD wIrIvmeH chut tutlh /

ml'QeD wIlaghmeH chut tutlh

canonical decomposition kanonische Zerlegung

	pini kulupu end of multiplication, product	ml'QeD vIqraq Ende der Multiplikation, Produkt

	pini kulupu insa inner product	ml'QeD vIqraq qoD inneres Produkt

	pini kulupu selo outer product	ml'QeD vIqraq Hur äußeres Produkt

	pini lili end of subtraction, difference	ml'QeD 'olQan / chuq Ende der Subtraktion, Differenz

	pini mute end of addition, sum	boq Ende der Addition, Summe

	pini pona result, solution	gher'ID Ergebnis, Lösung

	poka side	reD Seite

	sama to result in, to be identical to	chen / nIb ergeben, gleich sein

⊓⊓	**selo** outer form, geometric shape	tu'qom äußere Form, geometrische Figur
⊓⊓└∿‖↑	**selo pi linja tu wan** form of three lines, triangle	ra'Duch Form aus drei Linien, Dreieck
⊓⊓└∿↑=	**selo pi linja nasin sama** loS reD mey' Don form of parallel lines, parallelogram	Form aus parallelen Linien, Parallelogramm
⊓⊓└∿↑┤	**selo pi linja nasin taso** letbaQ form of orthogonal lines, rectangle	Form aus senkrechten Linien, Rechteck
⊓⊓	**sijelo** body, structure	yergh / Qur Körper, Struktur
⊓⊓ʊ	**sijelo musi** complex structure	Qur Qatlh komplexe Struktur
⊓⊓└∿↑=	**sijelo pi linja nasin sama** jav reD yergh Don structure of parallel lines, parallelepiped	Körper aus parallelen Linien, Parallelepiped, Spat
⊓⊓└∿↑┤	**sijelo pi linja nasin taso** letbaQ tut right parallelepiped rectangular parallel- epiped	Quader, rechtwinkli- ges Parallelepiped
⅃•	**sinpin** face	reD / yergh reD Seitenfläche

⊡	**sitelen** symbol, sign to depict, to write	maQ/Degh/ghItlh Symbol, Zeichen darstellen, schreiben
⊡ ⅏	**sitelen jasima** minus sign	maQ Dop / Degh Dop Minuszeichen
⊡ ∩	**sitelen nena** wedge	ley' qIv Keil
⊡ ◎	**sitelen sike** (big) dot	'oDtu' rutlh (fetter) runder Punkt
⊡ ᕲ	**sitelen wawa** boldface	ngutlhmey pl' Fettdruck
̈☐ #	**sona nanpa** mathematics	mI'QeD Mathematik
̈☐ I	**sona pini** conclusion	gher'ID ngaD Schlussfolgerung
V	**suli** large, long, length	tIn / tlq / 'abwl' / yamtaw lo'laHghach groß, lang, Länge
V ⊕	**suli ma** surface area	menwl' / pa' qoD lo'laHghach Flächeninhalt
V O	**suli ijo** volume	muqwl' / yergh lo'laHghach Rauminhalt, Volumen
̈○ L ̈☐ #	**toki pi sona nanpa** equation	wItte' Gleichung

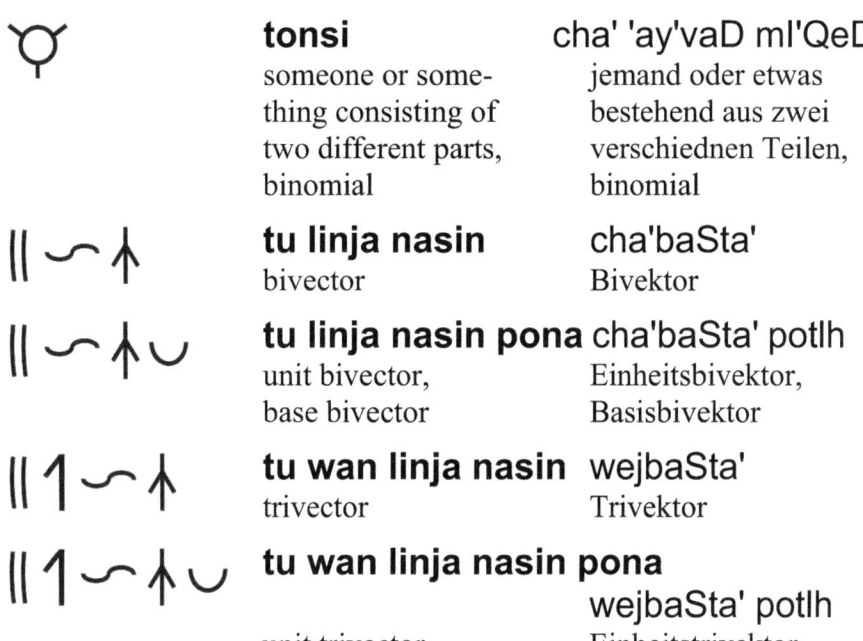

tonsi — cha' 'ay'vaD mI'QeD
someone or something consisting of two different parts, binomial — jemand oder etwas bestehend aus zwei verschiednen Teilen, binomial

tu linja nasin — cha'baSta'
bivector — Bivektor

tu linja nasin pona — cha'baSta' potlh
unit bivector, base bivector — Einheitsbivektor, Basisbivektor

tu wan linja nasin — wejbaSta'
trivector — Trivektor

tu wan linja nasin pona — wejbaSta' potlh
unit trivector, base trivector — Einheitstrivektor, Basistrivektor

nimi / pongmey

nimi ni sin en pali sitelen ni ona li ken ante:
choHlaH jIyweSmeyvam, mu'mey ghItlhlu'bogh je:

jan Kamen — qa'mer
Cramer / Er überrascht den Geist / He surprises the spirit

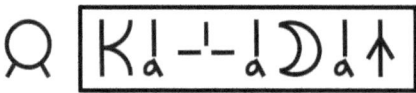

jan Kasaman loDnII

Grassmann / grassy man

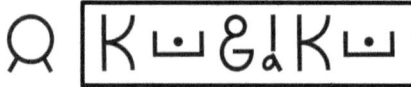

jan Kijaki DI'raq

Dirac / Schaf / sheep

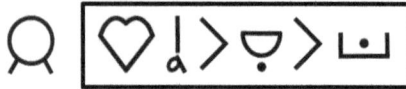

jan Paluli paw'll'

Pauli / Er ist am zusammenstoßen / He is colliding

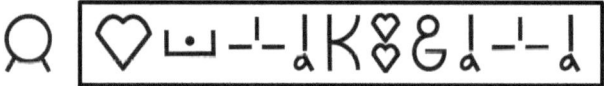

jan Pisakojasa peytlharghngongraS / pI'ta'qu'raS

Pythagoras / Der Tisch des Experiments erforscht die Säure / The table of the experiment explores the acid / Der Tisch, der fett sein sollte, ist wahrlich fett

jan Masin Elki mI'mey Dop Qaw'wI'

Martin Erik / destroyer of negative numbers

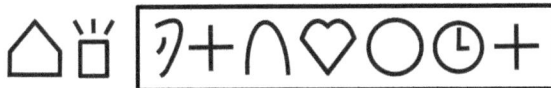

tomo sona Kenpite

qam blQtlq QI retlhDaq DuSaQ quv

University of Cambridge / Universität Cambridge / Ehrenhafte Schule
neben der Brücke am Fluss des Fusses / Honourable school at the
bridge of the river of the foot

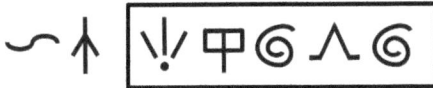

linja nasin Sinan   tlhIngan baSta'mey

klingonische Vektoren / klingon vectors

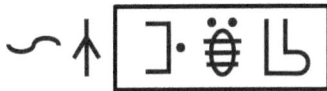

linja nasin San   chan baSta' tlq

Vektor in östlicher Richtung (90° ☀)
vector into eastern direction (90° ☀)

linja nasin Sinje   tIng baSta'

Vektor in südweslicher Richtung (220° ☀)
vector into south western direction (220° ☀)

linja nasin Ewe   'ev baSta'

Vektor in nordwestlicher Richtung (– 40° = 320° ☀)
vector into north western direction (– 40° = 320° ☀)

sitelen / ngutlhmey

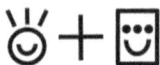

mi kepeken e sitelen kama pi ilo waso:
'orwl'pu' ngutlhmey vl'lo. naDev vl'agh:

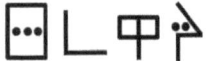

toki pona en sitelen pona
to'qlpo'na Hol, Sl'telenpo'na Hol je

sitelen pi ilo waso   **sitelen Elena**
'orwl'pu' ngutlhmey   'elaDya'ngan ngutlhmey

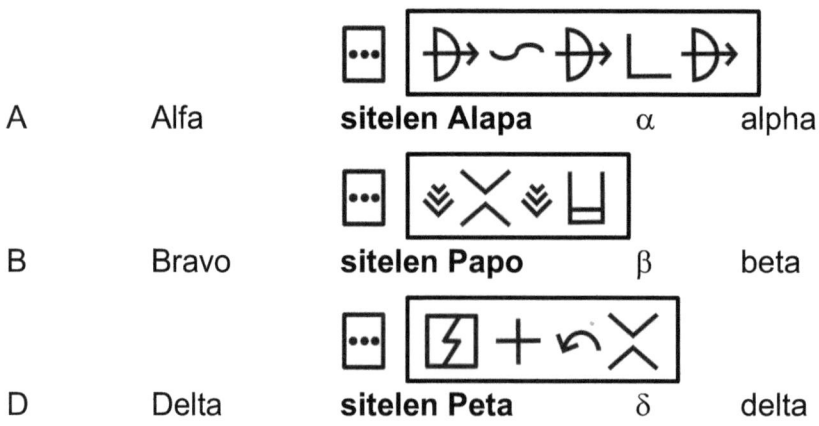

A	Alfa	**sitelen Alapa**	α	alpha
B	Bravo	**sitelen Papo**	β	beta
D	Delta	**sitelen Peta**	δ	delta

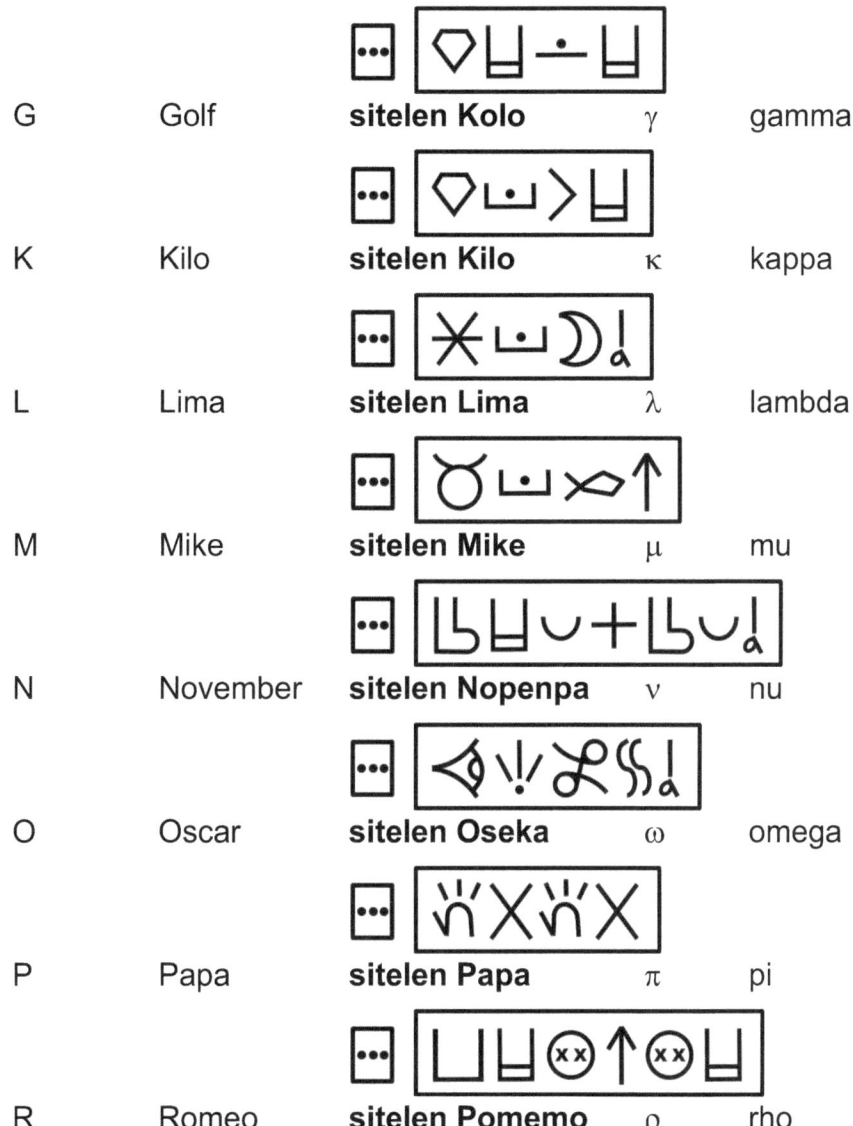

G	Golf	**sitelen Kolo**	γ	gamma
K	Kilo	**sitelen Kilo**	κ	kappa
L	Lima	**sitelen Lima**	λ	lambda
M	Mike	**sitelen Mike**	μ	mu
N	November	**sitelen Nopenpa**	ν	nu
O	Oscar	**sitelen Oseka**	ω	omega
P	Papa	**sitelen Papa**	π	pi
R	Romeo	**sitelen Pomemo**	ρ	rho

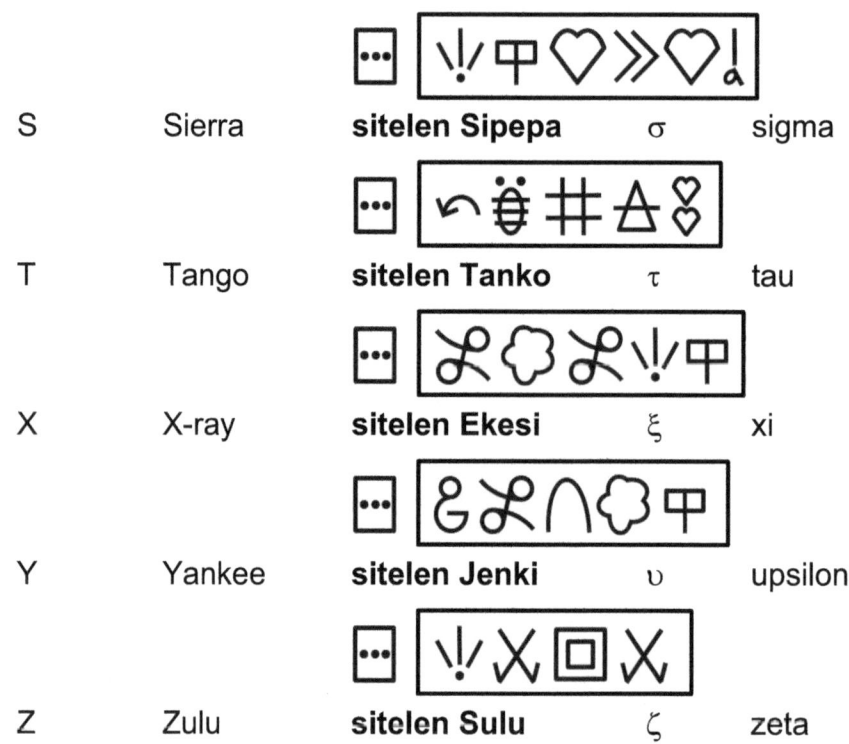

S	Sierra	**sitelen Sipepa**	σ	sigma
T	Tango	**sitelen Tanko**	τ	tau
X	X-ray	**sitelen Ekesi**	ξ	xi
Y	Yankee	**sitelen Jenki**	υ	upsilon
Z	Zulu	**sitelen Sulu**	ζ	zeta

mi utala e nanpa jasima:
mI'mey Dop vI'Qaw':

Martin Erik Horn: thcin se tbig nelhaZ evitageN.
Negative Zahlen gibt es nicht.
BoD, Norderstedt 2022 (ISBN: 978-3-7562-3808-8).

Martin Erik Horn: tsixE toN oD srebmuN evitageN.
Negative numbers do not exist.
BoD, Norderstedt 2022 (ISBN: 978-3-7562-5832-1).